The Emergence of Spacetime in String Theory

The nature of space and time is one of the most fascinating and fundamental philosophical issues which presently engages at the deepest level with physics. During the last 30 years, this notion has been object of an intense critical review in the light of new scientific theories which try to combine the principles of both general relativity and quantum theory – called theories of quantum gravity. This book considers the way string theory shapes its own account of spacetime disappearance from the fundamental level.

Tiziana Vistarini is a Visiting Scholar at the University of Rome, "Roma Tre", Philosophy Department.

Routledge Studies in the Philosophy of Mathematics and Physics
Edited by Elaine Landry, University of California, Davis, USA and Dean Rickles, University of Sydney, Australia

1 **The Logical Foundation of Scientific Theories**
Languages, Structures, and Models
Décio Krause and Jonas R. B. Arenhart

2 **Einstein, Tagore and the Nature of Reality**
Edited by Partha Ghose

3 **A Minimalist Ontology of the Natural World**
Michael Esfeld and Dirk-André Deckert

4 **Naturalizing Logico-Mathematical Knowledge**
Approaches from Philosophy, Psychology and Cognitive Science
Edited by Sorin Bangu

5 **The Emergence of Spacetime in String Theory**
Tiziana Vistarini

For more information about this series, please visit: https://www.routledge.com/Routledge-Studies-in-the-Philosophy-of-Mathematics-and-Physics/book-series/PMP

The Emergence of Spacetime in String Theory

Tiziana Vistarini

LONDON AND NEW YORK

First published 2019 by Routledge

2 Park Square, Milton Park, Abingdon, Oxon OX14 4RN
605 Third Avenue, New York, NY 10017

Routledge is an imprint of the Taylor & Francis Group, an informa business

First issued in paperback 2021

Publisher's Note

The publisher has gone to great lengths to ensure the quality of this reprint but points out that some imperfections in the original copies may be apparent.

Library of Congress Cataloging-in-Publication Data
Names: Vistarini, Tiziana, author.
Title: The emergence of spacetime in string theory / Tiziana Vistarini.
Description: New York : Taylor & Francis, 2019. | Series: Routledge studies in the philosophy of science ; 5 | Includes bibliographical references and index.
Identifiers: LCCN 2019006697 | ISBN 9781848935938 (hardback)
Subjects: LCSH: Space and time. | String models.
Classification: LCC QC173.59.S65 V52485 2019 | DDC 530.11—dc23
LC record available at https://lccn.loc.gov/2019006697

ISBN: 978-1-8489-3593-8 (hbk)
ISBN: 978-1-03-217792-2 (pbk)
DOI: 10.4324/9781315544151

Typeset in Sabon
by codeMantra

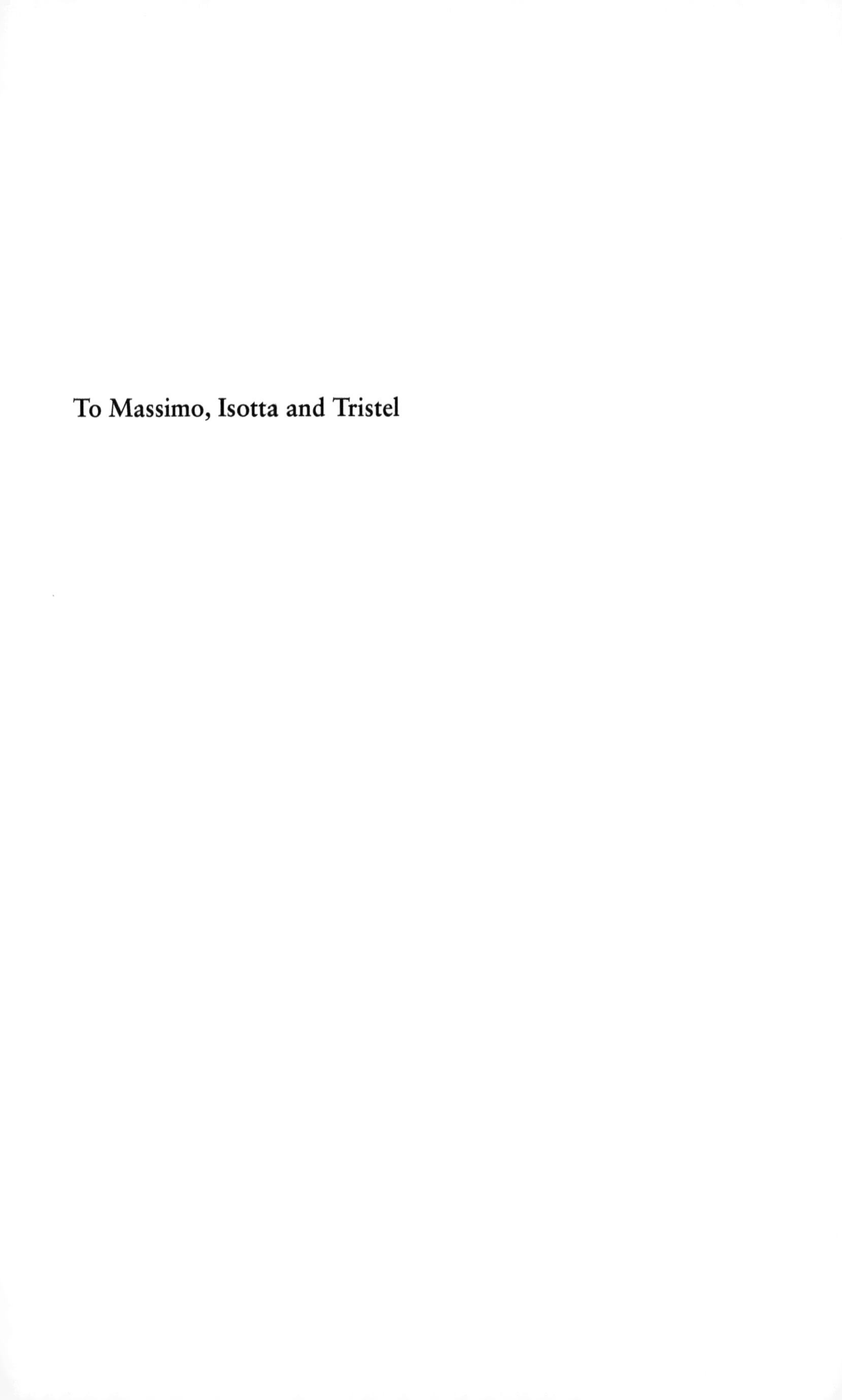

To Massimo, Isotta and Tristel

Contents

Introduction

It is widely held inside the quantum gravity circles that string theory is a background-dependent theory (namely a theory whose physical content is spacetime dependent) and that for this reason any attempt of tracing in the theory any notion of space and time emergence is simply a non-starter. In this book, I will develop a train of thoughts supporting a different, perhaps less popular view, namely, that string theory actually shows background independence in different ways and degrees. In order to rescue the theory from the charge of being background dependent, I will present a philosophical analysis of its physical content and formal articulation. I will argue that the theory delivers a quite composite notion of spacetime emergence and that this notion produces new metaphysical insights adding an important dimension on the traditional debate about emergence. A main metaphysical payoff is the natural consequence of the fact that in order to understand spacetime emergence in string theory, one is somehow forced to formulate some criterion of metaphysical explanation applicable to fundamental physics. This introduction contains a broad presentation of the book's structure. Each chapter analyzes the emergence of space and time in string theory from a different angle, but they all contribute to the same philosophical enterprise of producing a metaphysics sensitive to the physics occurring at the Planck scale.

Chapter 1 contains an introduction to the philosophical debate on emergence in general. It also describes the argumentative line intersecting any chapter along which my analysis of spacetime emergence in string theory develops. Finally, it illustrates how this analysis of spacetime emergence is complementary to the search for some non-causal metaphysical scheme of explanation in fundamental physics.

Chapter 2 is a self-contained presentation of some basic formal features of string theory. It only refers to the bosonic case. The bosonic string is a toy theory, but it presents the main physical insights of advanced versions of the theory in a relatively simple fashion. The goal is that of delivering a basic description of quantum strings physics that can be grasped without having a technical background and be useful to have a sense of the technical aspects of the debate on spacetime emergence in the theory.

Chapter 3 analyzes a crucial formal derivation (originally found in the early 1990s) of general relativistic spacetime structure from the quantum string dynamical structure. String theorists, along with many philosophers of the quantum gravity circles, tend to remain unimpressed by this derivation. It is widely held that the derivation is purely formal and background dependent. In this chapter, I argue against this view (see also Huggett and Vistarini, 2015). More precisely, I reconstruct the derivation, and I argue that it accounts for general relativistic spacetime emergence. The physical meaning of this derivation defines the status of general relativistic spacetime from the perspective of quantum string laws: general relativistic spacetime is an emergent byproduct of underlying quantum string dynamics that do not posit any fundamental geometry. This type of emergence produces a way of reading background independence through the perturbative formulation of string theory. Interestingly, this type of emergence also turns out to be interpretable via some non-causal metaphysical scheme of explanation.

Chapter 4 develops the idea of non-fundamentality of spacetime following a slightly different argumentative line. The methodology used to show the emergent nature of spacetime only involves the spatial extra dimensions introduced by the theory, and it develops around the notion of string dualities. The duality chosen in this chapter is T-duality.

Chapter 5 analyzes a physical scenario of spacetime non-fundamentality more radical than T-duality. Indeed, it analyzes the anti-de Sitter (AdS)/conformal field theory (CFT) duality. Also called holographic duality, this correspondence ties together two quite different physical worlds. It amounts to be a physical equivalence between a string theory describing a world with gravity (an AdS spacetime) and a conformal field theory living on its lower dimensional boundary, which does not contain gravity (namely, a Minkowski spacetime) – note that both AdS and Minkowski spacetimes are physical solutions of general relativity equations. This correspondence is the most concrete scenario in which some spatial dimensions of some bulk spacetime arise from dynamics occurring on its distant boundary. Those spatial dimensions are encoded in the boundary as internal degrees of freedom of quantum fields. Differently from T-duality, the non-fundamentality of spacetime here also points to the non-fundamentality of quantum string theory itself in virtue of its physical equivalence to a quantum field theory not containing strings. So, the holographic duality produces reflections on the emergent nature of spacetime that add some new philosophical dimension and physical meaning to the scenario produced by T-duality. In this chapter, I revise the analysis developed by me elsewhere about emergence in the AdS/CFT duality and in the de Sitter (dS)/CFT conjecture (Vistarini, 2016).

Chapter 6 formulates string theory background independence through the theory's moduli space. The local structure I posit on this abstract space unveils the different degrees of background independence characterizing the physical content of the theory. For this reason, this chapter presents a sort of unifying framework in which separate argumentative lines connected to different dualities can eventually join. As we will see, the structure I posit on this abstract space and the use I make of it do not completely overlap with the mainstream uses made within string theory's circles.

Chapter 7 concludes the analyzes of spacetime emergence in string theory by using a third methodology, namely, non-commutative geometry. Indeed, the chapter explores a possible way for getting some set of fundamental equations for some topological, non-geometrical structure in string theory. Non-commutative "spacetimes" are not really spacetimes; rather, they exemplify cases of topological backgrounds which are not geometrical since on them ordinary notions of duration and distance break down. However, ordinary spacetime (cosmological solutions of general relativity) is derivable from these exotic structures via low energy limits taken on their physical parameters. This fact, along with some theoretical findings about the endurance of quantum string theory when time exhibits a non-commutative behavior, is indicative of the theory's potential to account for deeper dynamics possibly happening in fundamental topological backgrounds like these ones.

1 Spacetime emergence and metaphysical explanation in fundamental physics

1.1 Emergence in what sense? Finding some bearings in a variegated philosophical landscape

Emergence is a controversial philosophical topic, and the landscape of its applications in philosophy of science is huge. Here, I neither consider the history of the concept, nor I reconstruct the debate exhaustively; rather, I refer to few of its contemporary uses which are relevant to us. Broadly speaking, in some contexts, emergence is used as a notion deeply intertwined with that of physical complexity. In some others, it constitutes the main core of some notion of incommensurability. Also, sometimes what emergence really means is actually supervenience. Within this variegated debate, one might individuate two main philosophical tendencies, each internally articulated.

Some scholars suggest that emergence is no more than an epistemological concept, that is, some sort of metaphor useful to grasp macro patterns resulting from micro interactions. The epistemological approach to emergence can be divided mainly into two schools of thought.[1]

The first is commonly known as the predictive interpretation. Emergent properties in this case are complex features of systems which cannot be predicted in virtue of our knowledge about the systems' pre-emergent stage. This notion of emergence is a diachronic one, and it delivers a scenario in which even the most accurate state of knowledge about the simple constituents' behaviors of the system cannot manage to completely predict its temporal evolution as a whole. One of the oldest formulations of this approach is that of Popper and Eccles (1977), although their thesis has been often read as stronger than a purely epistemological characterization of emergence. They identify emergence mainly to some degree of failure of causal determinism. In this setting, the emergence of hierarchical levels also appears to depend on an intrinsic fundamental indeterminism of the physical universe, and it seems to leave some room for downward causation, hence acquiring some degree of ontological robustness. In the same spectrum of prediction-based approaches to emergence, there is that of Bedau (1997).[2] His proposal delivers an epistemological notion of emergence not having the same hybrid nuances that make Popper's approach difficult to identify as purely epistemological. According to Bedau, an emergent state of any system is a macroscopic state in principle reducible to the system's underlying micro states, so there is no room for downward causation. This reducibility is a function of the amount of knowledge one might in principle gain about that macro state. The knowledge of any macroscopic (or emergent) state of any system can be in principle derived from the knowledge of the system's micro dynamics and of conditions external to the system. However, in practice there are some obstacles along the way of this reduction. The derivation at some point in time t_i of a complete knowledge of a system macro state could only be achieved by simulating all the microscopic interactions that through time – from the past initial conditions of the system to its state at time t_i – have been producing that emergent macro state. An emergent state in this case is seen as a property of systems exhibiting long-range behaviors. The latter are also characterized by a high sensitivity to small changes of the system's initial conditions. Indeed, the dynamics are not linear. So, unpredictability here is a de facto feature, and it is mainly due to the fact that fundamental physical properties can be specified only approximately by experimental methods.

Although interesting, a diachronic notion of emergence does not seem to play any role in the particular context of analysis developed in this book. Spacetime emergence in string theory does not really fit a scenario populated by phenomena whose emergent nature crucially unfolds through time, only because time is part of the emergent structure rather than part of some prior arena in which emergence unveils.

The second school of thought about epistemological emergence is commonly called the Irreducible-Pattern approach. Some of the views within this philosophical landscape turn out to play some role in my analysis of spacetime emergence. However, I argue that such a role is quite limited mainly because the epistemological character of spacetime emergence in string theory does not trivially fit these traditional schools of thought when string dualities are involved – as we will see in Chapters 4 and 5.

Within the Irreducible-Pattern approach, emergent properties are features of complex systems governed by laws within some special science. In this case, emergence is a synchronic property and it develops around the notion of irreducibility, rather than of unpredictability. Jerry Fodor (1974) is definitely an iconic expression of the Irreducible-Pattern approach: special science laws are irreducible to fundamental physical laws because of conceptual and "linguistic" obstacles; the macroscopic patterns described by the content of special sciences cannot be captured by the language and concepts of fundamental physics – this characterization will come back in Chapter 5 as I will be analyzing why this notion of epistemic emergence does not apply to holographic duality. Within the same spectrum of views, a quite innovative use of epistemological emergence – deeply influenced by the work of Nagel on the inter-theoretic account of reduction and emergence (Nagel, 1961) – is that by Batterman (2011). His work will be analyzed in Chapter 3. There, within a comparative analysis with the views of Butterfield (2011) and Kane (2013), I will explore their applicability to the emergence of general relativistic spacetime in string theory. The connection that Batterman establishes between emergence and inter-theoretic reduction branches from the central idea that the Nagel scheme of reduction of theories rarely applies smoothly. Indeed, the cases in which inter-theoretic relations via which the central concepts of the less fundamental theory are directly explainable in terms of the more fundamental one, are a minority. Batterman discusses many different physical phenomena arising at singular divergent limits for the relation of the two theories. The properties of systems arising at those limit values cannot be derived from the more fundamental theory. These are the properties that Batterman calls "emergent." So, emergence in this case is a failure of the reduction scheme. Against this robust notion of irreducibility, Butterfield argues for a different notion of epistemic emergence by identifying another way in which irreducibility might come into play. The singularity of the limit does not play any role in defining emergence because the novelty appears whether or not there is a singularity. The novelty always appears after the limit, either in case of smoothness or in case of re-normalized divergencies. The meaning of irreducibility here is close to the meaning originally proposed by Fodor. Irreducibility appears because of the epistemic incommensurability between the theories' languages and conceptual apparatuses. But these novelties preserve the reductive scheme. In this case, emergence is compatible with reduction. Note well that these approaches to emergence in terms of Irreducible-Pattern all share a central core idea: irreducibility involves the contents of the theories whose relation is analyzed – where content means physical content if the inter-theoretical relation studied is between physical theories, or more generally scientific content (or subject matter) if special sciences are involved. As I will argue in Chapter 5, this fact is exactly what makes this notion of irreducibility not applicable to spacetime emergence in string theory when dualities are involved.

Outside the debate on epistemological emergence, emergence is considered to be some robust ontological property. Traditionally, supervenience-based emergence is the received picture of ontological emergence. I will analyze supervenience emergentism in Chapter 3 in order to study its applicability to general relativistic spacetime emergence in string theory. Also in Chapter 5, I will refer to its iconic formulation by McLaughlin (1997) to explore what applicability it might have in the context of holographic duality.

However, it is not clear to me why supervenience emergentism is traditionally considered to be ontological emergence. As far as I know, this identification is puzzling for many. Indeed, already Kim (2006) pointed at the apparent tension between supervenience-based emergentism and the core requirements that ontological robust emergence should satisfy. For example, it is not clear how supervenience-based emergence accommodates for downward causal powers of emergent properties on the basal ones (also Wong, 2010). The distinction is so blurred indeed that one may say that the analysis of ontological emergence via supervenience might end up with being another form of epistemological emergence. My analysis of general relativistic spacetime emergence is an example of how this transition may occur (Chapter 3). General relativistic spacetime emergence admitted by string theory fits some conceptual features of supervenience-based emergentism. But in that case, the supervenient novelty is identified by mainly using epistemological criteria. In this case, one might say that emergence is compatible with physicalism.

Now, the general approach underlying my use of emergence might be seen as a form of metaphysical foundationalism. One might start saying that physical reality comes into layers: without specifying for now what "physical fact" stands for, every physical fact is either fundamental or derived with respect to more fundamental physical facts. As we will see in the next section, the notion of grounding – or at least my interpretation of that notion – plays here an essential role. In particular, every physical fact either is grounded in physical facts that themselves aren't grounded or it isn't grounded. Then, the physical world is hierarchically structured from physical fundamental facts to increasingly complex derived facts.

Physical facts that aren't grounded are those defining the fundamental physical layer at the bottom of the hierarchic structure, that is, they are the physics occurring around the smallest length scale admitted by quantum gravity approaches like string theory and loop quantum gravity. Fundamental physical facts are physical properties and dynamics predicted by fundamental laws – laws applicable to the smallest length scale. Derived physical facts, informationally more complex than the basal ones, are in turn defined by larger length scales in spacetime and predicted by physical laws which are lower energy limits of the fundamental ones. Each derived layer contains novel physical qualities with respect to those of its basal layer. The novelty is an entirely new type of property, and it emerges synchronically from the basal properties.

So, at each stage of the hierarchy, the emergent laws are derived from the lower-level ones by means of physical limits taken on the lower dynamics. The limit can be singular, and for those cases in which the divergences cannot be eliminated, the emergent laws constitute some robust supervenient novelty. Instead, for those cases in which the divergences can be eliminated, the emergent laws constitute some weaker supervenient novelty in virtue of the well-defined existence of the limit.

However, even in this case, the supervenience of the higher laws on the lower ones still brings in a notion of deep incommensurability between physical scenarios because of the existence of the physical limit that hardly fits traditional reduction schemes.

1.2 Spacetime emergence and non-causal metaphysical explanation in string physics

Now that we found our bearings within the philosophical landscape, I would like to start with unpacking my specific use of emergence. As I said in the introduction, emergence here applies to spacetime and from the perspective of the fundamental physical ontology of string theory. This specification brings in almost instantaneously two sources of philosophical perplexity. First, a still unspecified notion of non-fundamentality applied to features of the physical world traditionally considered to be constitutive of some fundamental arena in which physical processes unfold. Second, the perspective from which the emergence is analyzed is that of a physical theory whose physical content is not directly accessible. Both concerns are the topics of the next two sections, respectively. In the first, I argue that the non-fundamentality of spacetime – as it appears from the perspective of quantum gravity laws – also partially appears from the perspective of classical physical laws, at least according to my relational reading of the latter. In the second, I argue that the lack of direct empirical access to the high-energy physical content of string theory does not undermine its role of adequate "empirical source", or by using a popular van Fraassen's expression, it does not undermine the fact that the theory can save the phenomena (van Fraassen, 1980, p. 12).

Now, explaining the emergent nature of spacetime in string theory is a composite task. One first needs to look at the physical aspects of this emergence, that is, the theory's physical content, its invariance via dualities that do not preserve geometry, and its low-energy behaviors. Then, one should philosophically understand what kind of explanation all these things constitute.

Let's start with arguing why traditional mechanical explanation would not apply to spacetime emergence. Looking at the historical use of this explanatory scheme might contribute to make the point. The vocabulary in which mechanical explanations of phenomena are usually formulated presupposes at least time as fundamental entity, and frequently time-directness on top of time, because of the fundamental role assigned to causality by that vocabulary. Note well that this point is not merely linguistic. It indeed reveals the substantive role played by the notion of initial condition within any school of thoughts about mechanical explanation. Any arbitrarily chosen initial condition crucially accounts for the system's behavior (in combination with the corresponding dynamics). The explanatory role of an initial condition is in virtue of its being an initial physical state of the system at some point in time and in space. After all, this is the great legacy of modern scientific revolution. Historically, the mechanical explanation of phenomena was proposed in opposition to the teleological one, according to which explanations of phenomena may reside outside the *res extensa*. By claiming that space and time are emergent because they are physical features susceptible of mechanical explanation, we are also simultaneously depriving them of the role of *explanans* they both have within the explanatory scheme. It is hard to imagine what is left of the mechanical explanation once space and time are removed from the list of its *explanans* constituents. Things get worse if one includes in this analysis the causal reasoning traditionally built into the mechanical explanatory scheme. Causal explanation requires that the *explanans* and the *explanandum* both reside in the same physical spacetime. The foundational issues affecting any quantum gravity approach, including string theory, mainly arise from the combination of two factors – on the one side, the difficulty in applying traditional causal explanation to spacetime emergence and on the other, the difficulty in finding metaphysical schemes of explanation that are non-causal and smoothly fit their physical content. In this section, I make an attempt of at least identifying some plausible candidate that might fit string theory's physical content.

Now, spacetime emergence in string theory might be read through some metaphysical explanatory scheme deeper than the traditional causal one, still reproducing the latter. In this sense, traditional causality would be the emergent explanatory structure of the emergent classical physics. A non-causal metaphysical explanation fitting string theory might be found by combining two schemes: the unificatory and the grounding schemes. Their appropriateness appears to be supported by the theory's physical content. Let's unpack this point by starting to illustrate the use I make of these two schemes and how they both relate to the notion of explanation.

The fact that the theory admits a minimal length in spacetime contributes positively to the attempt of combining its physical content with the grounding scheme. In this hybrid metaphysical scenario, the most fundamental layer of the hierarchic structure would be defined by this minimal length. One might say that the characterization of layers mainly in terms of physical length scales produces a "naturalized" form of metaphysical foundationalism.

Now, in what sense grounding is an explanatory relation in general? The characterization formulated by Fine in purely metaphysical terms captures quite well this general sense: "if the truth that P is grounded in other truths, then they account for its truth; P's being the case holds in virtue of the other truths' being the case" ("The Question of Realism," Philosophers' Imprint, 2001).

Now, is there any broad logical form of general grounding statements accounting for this relation? Broadly speaking, there are two types of answer. The first type basically says that if I claim that I can vote in Italy because I am an Italian citizen, I am formulating a grounding statement with a non-truth-functional sentential connective. In other words, supporters of this position recognize the explanatory scheme and at the same time remain neutral about the existence of the facts that I can vote in Italy and I am an Italian citizen, and finally they are neutral about the grounding relation holding between these two facts (Fine, 2001). The second type of answers commits to the existence of grounding relations, that is, it considers grounding statements as claims which are structured by truth-functional sentential connectives. In general, these connectives are truth-functional relational predicates having the canonical form of binary relations between facts or groups of facts (Audi, 2012; Rosen, 2010).

Now, in the specifics of my application of the grounding relation as explanatory scheme, I endorse the idea of its existence, but I revise its purely logical formulation in light of the formal language built into string dynamical models. What I mean is that the grounding relation exists in virtue of a physical limit connecting the more fundamental physical parameters of the string laws to the physical parameters and laws of a derived, emergent spacetime. The language formalizing this grounding relation is imported from functional analysis, rather than predicates logic. The analysis of general relativistic spacetime emergence in Chapter 3 exemplifies a particular application of this revised grounding scheme. In that case, one has a spaceless and timeless string dynamical model, that is, a set of fundamental laws at the smallest length scale, whose low energy limit produces emergent general relativistic spacetime, along with the emergent laws predicting those sets of geometries.

Now, one cannot really say that the fundamental string laws causally explain spacetime emergence because according to the theory, those fundamental dynamics are outside spacetime. One might instead say that the fundamental string laws explain the emergence in virtue of the fact that they ground that emergence, that is, in virtue of the fact that they are a sort of condition of possibility for that emergence to occur. So, string laws would explain spacetime emergence virtually in the same way in which my Italian citizenship makes it possible for me to vote in Italy.

As I said above, the unificatory scheme is the other non-causal metaphysical scheme of explanation that in the case of spacetime emergence in string theory smoothly combines with the grounding scheme. Its appropriateness is in virtue of the fact that string theory's physical content delivers a unifying framework explaining all sorts of physical interactions. Some crucial structural features of the underlying string dynamics can reproduce at low-energy classical gravity on one side – classical in the sense of general relativity – and ordinary particles physics on the other. Indeed, string theory quantizes gravity by also simultaneously delivering a unifying physical ontology. The unificatory role becomes manifest via different physical limits taken on the same string dynamical structure. So, the unificatory scheme plays an explanatory role complementary to that of the grounding scheme. The latter explains what set of underlying physical conditions makes it possible for low-energy scenarios to look like they look, whereas the former unifies all of them by unveiling the shared set of underlying physical conditions making them possible.

1.3 No time for strings, nor space

This section starts by leaving aside for a moment string theory. A couple of general questions to start with are the following: "in what sense would space and time be susceptible of explanation?" and also "would they be susceptible of explanation in the same sense in which material features of the world like tables and chairs are?" What I mean with "material features of the world" is simply anything that relates to distribution of material bodies over some background, as opposed to those feature of the background itself conceived somehow prior to that material distribution. Also, in this context, something is susceptible of explanation just in case it is explainable in terms of distribution of material bodies and of the laws governing them.

It is widely argued inside many philosophy of science circles that the macro description of the world delivered by classical, pre-relativistic physics unequivocally assigns to space and time the status of background's features, somehow prior to the material ones. Before even mentioning the deeper quantum string perspective, I argue that this way of interpreting the lesson of classical physics about space and time is questionable. An alternative interpretation amounts to trace already in pre-relativistic, classical physics the idea that space and time are "derived" features of the world, rather than being an arena somehow prior to any material interactions unfolding in the universe. Indeed, any classical, pre-relativistic Hamiltonian potentially contains an embryonic notion of spacetime emergence, a sort of classical, pre-relativistic counterpart of the finer quantum gravity notion. Here, I am pulling out this classical counterpart and showing how it can be seen in continuity with the quantum one delivered by string theory.[3]

Think of a pre-relativistic classical Hamiltonian describing the dynamics of some classical system of particles, moving around in a three-dimensional Euclidean space. The Hamiltonian of a system is a function from which we can extract exhaustive information on the physical law obeyed by the system. A pre-relativistic, classical Hamiltonian looks like

$$H = \sum_{i=1,..N} [KINETICPART] + \sum_{K=1,..,N-1} U_{kj}((x_k - x_j)^2 + (y_k - y_j)^2) + (z_k - z_j)^2),$$

where $k \neq j$.

For the purpose of this introduction, the kinetic part can be neglected. The x, y and z coordinates showing up in the function are spatial cartesian coordinates in a space equipped with a Euclidean metric. For now, let's leave aside the time coordinate.

The formal structure of the potential U remains unspecified in this formula. But we don't need any further detail. It suffices to say that any potential of any classical Hamiltonian is structured to accommodate all sorts of particle interactions. Also, as we can see in this formula, the potential is structured to show the Euclidean background geometry, posited as the actual geometry since the start.

Let's pretend for the sake of the argument that all we need to account for macroscopic systems like tables, chairs, cats and humans, and so on is the set of interactions described by this classical potential – that is, let's pretend we are inhabitants of a world which is fundamentally classical and in which there is no such a thing like quantum behaviors.

How would a world like this appear to us? Here, the expression "world's appearances" does not contain any reference to our subjective experience. Rather, it simply denotes what the world looks like through the outcomes of experiments and of measurements.

A physical world like this would appear, without any doubt, to be three-dimensional and Euclidean. But why exactly? A satisfactorily answer may be found by looking at the mathematical form of the potential U in the Hamiltonian. A salient feature of this potential is that of being explicit function of the three-dimensional Euclidean distances. Also, another crucial feature is that the potential dictates how any material interaction occurs.

So, one may say that in this case, material interactions make the Euclidean distances manifest. That means, the potential produces a manifest geometrical image of the world that in this case coincides with the actual one. But what if the relationship between the actual background geometry and the mathematical form of the Hamiltonian is more complicated than that? What if the manifest and the actual geometry do not coincide?

There is an old story about this possible geometrical mismatch that I illustrate in Chapter 4 in the context of string dualities. It is the parable by Poincaré about the difficulty of establishing beyond any doubt the status of geometrical knowledge. According to Poincaré, the issue cannot be settled by means of empirical evidence. The original context in which Poincaré delivered his parable was the one in which philosophers and mathematicians were debating inside the epistemology of geometry. In this introduction, I won't be mentioning the historical features of that debate. Indeed, the lesson I want to draw here is at right angle with all that. It is a lesson about physics.

The Poincaré's story about the impossibility to decide by means of empirical data what is the correct geometry of the world tells about an imaginary two-dimensional world equipped with Euclidean geometry. It is a disk, the one that contrives by means of effects of spatial variation of the temperature on the lengths of measuring rods. The inhabitants, unaware of the hidden dynamics affecting rods' lengths, get out of measurements and of dynamical generalization a manifest geometrical image of the world that is not the same as the actual one. They believe to live in an infinite Lobachevskian plan. In this case, one may say the dynamics are producing a manifest geometry of the world that does not coincide with the actual one, however deep is the detail of the inhabitants' empirical investigations.

What this parable suggests to me is that we may generalize the classical scenario described above to a case of arbitrarily curved background. In this case, the Hamiltonian would depend on generalized coordinates x, y and z. In a world having this space geometry, there is no general and unique way of defining the line element ds^2; there is no uniform way of defining the distance between particles (since the integral of the line element define distances). Nonetheless, in a Hamiltonian like this, the potential can appear (by means of local coordinates' transformations) to be the function of $(\delta s)^2 = (\delta x)^2 + (\delta y)^2 + (\delta z)^2$:

$$H = \sum_{i=1,..N} [\ldots] + \sum_{K=1,..,N-1} U_{kj}((\delta x)^2 + (\delta y)^2 + (\delta z)^2),$$

where $k \neq j$.

More generally, it means that interactions between particles can be seen as manifesting a Euclidean flat space geometry. And so, the world we are considering now, in spite of the fact that it is actually curved, appears to its inhabitants to be flat and Euclidean.

So, we might wonder what it means to say that space is actually curved in this case. The point is that in this new case, the actual curved geometry of the background appears to have no role in the production of the flat geometrical appearances. It is the dynamics that do all the work. In other words, the Hamiltonian produces the manifest image of a flat background in a way that is independent of the actual curved background. Therefore, one may also inevitably conclude that the Hamiltonian produces a manifest geometry of the world in a way that is independent of whether or not the background has any intrinsic geometry. We don't need to postulate any fundamental geometry to account for the manifest geometrical appearances of the world. Then, the moral of the story is that we can extrapolate from a pre-relativistic, classical Hamiltonian an idea of spacetime emergence. The manifest geometry of the world seems to have nothing to do with its actual structure; rather, it seems to be an emergent thing.

Now, the reasoning followed so far is generalizable to any quantum Hamiltonian. More importantly, it is exactly from a quantum perspective that the separation between manifest geometry of the world and actual fundamental structure becomes quite evident. In the quantum gravity description delivered by string theory, the manifest image of the world and the world's actual fundamental structure – as it is described by the theory – fall apart.

In "After Physics," David Albert develops a similar train of thoughts heading to the idea that the fundamental non-spacetime "arena" in which quantum dynamics unfold is the space of configuration of the universal wave function. If my reading is correct, the space of configuration in this scenario would amount to be the totality of all possible values taken by the universal wave function at any particular time. This manifold is not geometrical at all; rather, it is simply topological. The manifest geometrical image of the world, the one we are familiar with, would emerge somehow from this configuration space by means of the work made by the quantum mechanical Hamiltonian.

My analysis of spacetime emergence in string theory has some stylistic aspects in common with this way of pointing to the non-fundamentality of geometry. In Chapter 6, I argue in favor of string theory background independence by using an abstract space related to the theory and encoding the totality of possibilities. By exploring the local structure of this abstract space of possibilities, one can conclude that the fundamental features of the physical world should be sought primarily in the features of fundamental laws without positing any additional fundamental structure not contained in them.

Now, reiterating the initial question, can we say that the manifest geometry arises in the same way in which macroscopic things like tables and chairs emerge? I argue that for some aspects we can. However, for some others, we cannot. What I mean is that we can say that tables and chairs emerge from the collective behavior of microscopic particles guided by microscopic laws and that at that microscopic level, there aren't any things like tables and chairs. A similar claim virtually applies to the emergent nature of space and time. Based on the line of reasoning developed so far, spacetime appears to be a byproduct of underlying fundamental dynamics unfolding in some underlying structure that might not be spacetime at all. Simply postulating a topological structure might do all the work. However, the emergence of tables and chairs can be mechanically explained, whereas spacetime emergence cannot smoothly fit that explanatory scheme. As we saw in the previous section, a requirement for traditional mechanical explanation, shared by many different schools of thought about the scheme, is that the *explanans* and the *explanandum* should share the same spacetime. Otherwise, any talk of causality naturally encompassed in the mechanical scheme of explanation would hardly apply. So, understanding the sense in which space and time are susceptible of explanation requires an extension of the traditional mechanical sense. In the previous section, the combination of the grounding scheme with the unificatory one is tentatively used to serve this purpose. The fundamental spaceless and timeless string laws constitute the *explanans* outside spacetime that would account for spacetime emergence not in causal terms but as a condition of possibility for that emergence to occur. Nothing beyond the fundamental laws needs to be posited at the fundamental level in order to explain the manifest image of the world. Even if something is posited, it would not have any explanatory role.

Now, positing at the fundamental-level phenomena's features irreducible to the fundamental laws is a quite common philosophical strategy used to account for ordinary phenomena that don't get reflected directly in those laws. But one might claim that this strategy does not really achieve the goal. Indeed, there isn't any clear sense in which positing something that resembles phenomena can play any role in explaining them. The logic underlying this strategy appears to be close to the logical fallacy of begging the question. Then, for the sake of producing "empirically adequate" metaphysics fitting the physics at the Planck scale, we cannot overlook the fact that all quantum gravity proposals (although very different) share the same insight that only the fundamental laws (whatever they are in each account) must be in charge of explaining the manifest image of the world. And we cannot ignore the fact that they also share the idea that spacetime – far from being a background feature prior to those dynamics – is an emergent feature populating the manifest image of the world.

Now, the claim that space and time are emergent does not entail – at least according to string theory – that space and time are not real. Indeed, I want to emphasize that the idea I am opposing in this book is not that space and time are real. Rather, by endorsing string theory perspective on their physical nature, I argue against the metaphysical idea that they are built into the fundamental structure of the physical world. Space and time are real features of the world, but they are derived, emergent structures. Of course, the resources being presupposed by ordinary linguistic locution and ordinary mechanical explanation had better be real and locatable in the physical world. But there is nothing about a denial that space and time are fundamental that carries with it a denial that these things are not in the physical world. My thesis is that spacetime can be located in the world like derived structures. There is nothing in the profusion of our language (both ordinary and scientific) that the things at which it points to must be fundamental rather than derived. My game in this book is still that of saving the phenomena.

As we will see, there are more than one way in which string theory delivers a physical scenario in which space and time are emergent. Quantum string dynamics show independence from geometry in many different ways. The most radical case of independence is that in which string laws appear to be the condition of possibility for geometry to exist. Then, it is in virtue of its background independence that the theory must face the same challenge as every other quantum gravity proposal, namely, that of building an explanatory bridge between the quantum string scenario and the manifest one. Filling the explanatory gap between fundamental laws and phenomena not reflected in those laws is still an open issue for philosophical approaches, like mine, willing to produce a metaphysics sensitive to the sources of fundamental physics.

However, more traditional approaches to metaphysics claiming its independence from fundamental physics would not serve the purpose better. Indeed, any metaphysical posit about the fundamental existence of space and time, postulated to explain our ordinary experience of space and time, would not fill the gap. It's not clear at all what conceptual category can describe in this case the relation between the *explanans* and the *explanandum*. While there is a clear sense in which we can meaningfully say that the presence of a lamp on my desk "causes" my impression that there is a lamp on my desk, there isn't any clear sense in which we can state that positing a fundamental spacetime geometry (in addition to the fundamental laws) may explain my impression that our world in its basics is in space and time.

1.4 Some source of trust in string theory

It is widely held within the philosophy of science circles that the great expectations originally harbored for string theory back in the mid-1980s have been downsized approximately ten years later for some presumed theoretical stumbling blocks preventing string theorists from making definite predictions. The obstacle would be the overabundance of possible shapes for extra dimensions, all compatible with the same low-energy physics. However, this feature might well be read in a completely different way, namely, as a positive symptom of the theory's background independence – this reading of the underdetermination problem in favor of the theory's background independence will be pushed forward in Chapters 4–6, where the theory's dualities are a main argumentative tool.

Although the extra dimensions are emergent in string theory (as I will argue), its equations place restrictions on what kind of geometrical shape they are supposed to have. They must belong to a specific set of shapes, also called the set of Calabi-Yau structures. However, this set is quite large and contains a huge number of different shapes, all Calabi-Yau.

Now, extra dimensions with different shapes are in principle supposed to produce different low-energy particle physics. But in string theory, it became clear soon that this expectation was not going to be met. Indeed, a main feature of string theory is exactly the underdetermination of its physical content by the Calabi-Yau geometrical shapes of its extra dimensions.

Now, if on the one side that makes mathematically intractable the problem of singling out a particular extra dimension shape as the one existing at the fundamental level, on the other side, it shows the non-fundamental nature of any geometry, including that of the extra dimensions. After all, the mathematical intractability of the problem of singling out the correct fundamental geometry of the extra dimensions might be simply due to the fact that this is the wrong problem. The lesson might well be that in the theory, any geometry cannot be considered fundamental because the fundamental structure of the world is not geometrical at all. So, people might well have been reading the underdetermination problem in string theory in the wrong way, undervaluing the theory's potential in terms of background independence. After all, the idea of extrapolating from the theory some information about the correct geometry of the world at the fundamental level undermines any talk of background independence since the start.

Another fact used against the scientific viability of string theory is the lack of directly testable predictions of its physical content, i.e., predictions that can be directly confronted with measurement and observations. Any confrontation with measurement and observations can only be done through the mediation of some low-energy theory. For this reason, many of its critics claim that string theory, as it presently stands, cannot avoid the charge of being divorced from genuine experimental evidence.[4]

Now, against this claim, one might argue that this appeal to direct empirical access is a form of orthodox verificationism smoothly applicable only to classical physics. Indeed, not even a theory like quantum mechanics – widely considered as one of the most empirically robust achievements of the 20th century – would pass the test of verificationism. Indeed, what makes quantum mechanics empirically robust by connecting its deterministic laws to the probabilistic outcomes of measurements is a twisted relation that one would never find in classical physics and that would never pass the test of direct empirical verification. This twisted relation is also well known in the philosophy circles as the measurement problem.[5]

Now, string theory is a quantum theory more fundamental than ordinary quantum mechanics in virtue of the fact that it includes quantized gravity in the picture. So, assuming that the measurement problem in ordinary quantum mechanics is due to incompleteness, string theory should in principle fix the measurement problem. But the presumed finer explanatory power comes with energy scales much higher than that reachable by presently available technology. So, one might double down against string theory's viability and conclude that the only reason for which there is no measurement problem in string theory is that there are no performable measurements.

Far from this conclusion, I argue that an alternative argumentative path might be that of revising the traditional criteria of verification. Indeed, one might work on finding a wider set of criteria encompassing the old traditional ones and also smoothly fitting the physical content of any existing quantum gravity approach. After all, the fact that methods of direct experimental verification are not smoothly applicable to a theory denying the fundamental existence of space and time is not very surprising: measurements and observations would not be performable outside space and time.

So, a key insight to broaden the traditional set of criteria might be that of giving centrality to the indirect link that quantum string theory has with physical phenomena. Indeed, the way in which string theory establishes contact with them also provides an easy way out from the charge of being divorced from experimental evidence – see also Huggett and Wuthrich (2013). Indeed, as I said earlier on, quantum string theory reproduces at low energy the empirical contents of two very different theories, both successfully tested, namely, general relativity and the Standard model of particles.

Now, in defense of the empirical coherence and adequateness of the theory, I will mention and reinforce the line of reasoning presented by Richard Dawid in his "String Theory and the Scientific Method" – see Dawid (2013). His argumentative strategies offer some robust source of trust in the theory viability, and they provide a possible way of broadening the set of traditional criteria, one that uses the conceptual novelties introduced by string theory in the traditional methodology of science. Without removing the role of empirical data as the ultimate judge of the theory's viability, the author points to those features of string theory that might be considered as the new set of requirements for scientific viability that broadens the traditional set. A main virtuous feature is the theory's unexpected explanatory coherence. This happens in physics when a new physical idea produces unexpected solutions for other problems that idea was not originally supposed to solve. And string theory over the years has been showing such unexpected explanatory coherence in more than one occasion. String theory leads to gravity, supersymmetry, black hole entropy, and so on.

Moreover, via Dawid's no-alternative argument, there aren't many viable unifying theories that can compete with the self-consistency and explanatory power of string theory. As I said earlier on, string theory quantizes gravity by also delivering a unifying framework for all the existing fundamental forces. Such unifying explanatory power should be added to the list of requirements for scientific viability. Finally, the last requirement for scientific viability of a theory amounts to be the experimental success of other theories within the same research program. String theory develops elementary particle physics, and it maintains a deep connection to its origin – Dawid's meta inductive argument. As far as I know, there is a concrete sense in which one can say that elementary particle physics can be derived from string theory; that is, some particular configurations of strings define string models of particles. The latter reproduce at lower-energy regimes the chiral fermions and gauge bosons of the Standard model. In this sense, the Standard model can be thought as an emergent, phenomenological theory, deduced within a certain approximation from string theory. The Standard model is a well-confirmed theory according to traditional criteria. So, if the emergent lower-energy theory works well, physicists should have some strong reasons to believe that the reducing higher-energy theory will work well too.

Notes

1 See also O'Connor, Timothy and Wong, Hong Yu, "Emergent Properties," The Stanford Encyclopedia of Philosophy – Summer 2015 Edition.
2 Mark Bedau, *Weak Emergence*, Noûs, Volume 31, Issue 11, 1997.
3 My line of reasoning here is similar to that developed by other authors. David Albert in his "After Physics" (Albert, 2016, ch. 1, 6) presents a similar argument in the context of quantum mechanics. Also, a similar line of reasoning is followed by Harvey Brown in his "Physical Relativity: Space-Time Structure from a Dynamical Perspective" (Harvey, 2005) and by Nick Huggett in his "Essay Review: Physical Relativity and Understanding Space-Time" (Huggett, 2009).
4 Parenthetically, having a physical content not directly accessible via measurements and observations is a feature that string theory shares with any other existing approach to quantum gravity, but this feature somehow only seems to undermine the scientific viability of string theory.
5 Here, I briefly mention just a main philosophical controversial feature of quantum mechanics. Given the deterministic nature of the quantum mechanical laws and of their theoretical predictions, repeated measurements of identically prepared systems, that is, in the same initial condition, should produce the same outcome. But this is not what happens, whether the same measurement is performed through time on identically prepared systems or it is performed simultaneously on a collection of them. The empirical corroboration of quantum mechanics' theoretical predictions does not come from a single measurement performed on a system, never mind how much sophisticated the measurement device is; rather, it can only come from a set of measurements that eventually reproduce the quantum probability distribution indicated by the system's wave function. Now, in addition to having this twisted connection to measurements, quantum probability, differently from the classical one, also shows interference patterns. In other words, any individual quantum system interferes with itself and evolves deterministically when it is not measured, whereas it behaves stochastically and without self-interference when a measurement is performed.

2 Strings in a nutshell

This chapter introduces the basics of string theory through consideration of the bosonic string. The latter is an unrealistic model, still it is instructive. In the absence of matter, the bosonic string is a relatively simple context to grasp the fundamental insights of the fully articulated theory. So, this chapter describes some basics of string theory's physical content and of its formal articulation. It also prepares the ground for extrapolating from them the theory's metaphysical lesson about the non-fundamentality of space and time.[1]

2.1 Some introductory remarks

The first attempt of unifying quantum theory and relativity was done by combining quantum mechanics and special relativity. The result, called quantum field theory, left gravity out of the picture. Then, it is not surprising that in quantum field theory spacetime is a fixed arena prior to the material processes unfolding in it. In other words, spacetime is not solution of the dynamical equations of quantum field theory.

Now, the notion of point in spacetime plays an important role in quantum field theory. As far as I know, the use of this notion within the framework does not entail or require any endorsement of spacetime substantivalism. Indeed, spacetime points are fundamentally a way of speaking about physical interactions. Spacetime points are brought into the framework because it is filled with fields, and each field produces (or destroys) its own quanta.[2] Loosely speaking, because of that, interactions in quantum field theory occur at zero distance. Particles produced by those interactions are *point particles*, individually located at a single spacetime point (see Srednicki, 2007, chapter 1).

Now, in quantum field theory information about particles interactions can only be gained by computing the probabilities for those interactions to occur. The formal language of these probabilistic predictions is that of perturbation theory. A perturbative expansion is simply a sum in which each term describes an interaction, each term showing a degree of physical complexity different from that shown by any other term. In other words, each term refers to the same interaction, having fixed initial and final states, but each term depicts what happens during the interaction with a degree of complexity different from any other term in the sum. Usually, the more terms you add to the sum, the higher is the internal complexity of the interaction one considers – a high degree of internal complexity is often formally encoded in a high polynomial degree. One might say that each term of the sum describes a possible physical way in which the interaction might unfold between its initial state and its final one. The degree of physical complexity of the unfolding interactive process depends on the degree by which one perturbates that process – perturbating some physical process basically means interacting with the process.

Now, this computational apparatus works quite well with electromagnetic interactions, weak and strong interactions. But it does not work with gravitational interactions. Why is that? For a long time, it has been known by physicists that gravitational waves carry angular momentum equal to 2. From this property, they deduced that the hypothetical quantum of the gravitational field would be a spin-2 particle.[3] The theory also predicts the semi-classical behavior of any of their coherent superpositions, which turns out to be exactly the behavior of gravity according to general relativity. As I said, any particles interaction in quantum field theory occurs at a single spacetime point; that is, it occurs at zero distance. This is the reason for which, regardless the type of interaction, calculations of interactions inevitably produce energy divergencies. The Heisenberg uncertainty principle explains this correlation. According to the popular principle, there is an inverse relation between the relative distances of particles positions and the variation of their energies, i.e.,

$$\Delta x \Delta p \sim \frac{\hbar}{2}.$$

If one considers a typical quantum field theory interaction, namely, an interaction at zero distance, the limit $\Delta x \to 0$ produces the divergent energy limit $\Delta p \to \infty$. This infinity produces the divergent calculations mentioned above. However, in any kind of interaction, except for the gravitational one, those divergencies are killed. Divergencies in electromagnetism, or in weak and strong interactions, can be renormalized. But those arising from gravitational interactions are not because of the combination of two factors: on the one side, the zero distance at which gravitons interact and, on the other, the gravitons spin number.

Now, string theory gets rid of zero-distance interactions since it replaces point particles with one-dimensional strings. So if one doesn't go all the way to $\Delta x \to 0$, then the limit is cut off at some distance, still small but not zero. That leads to non-divergent computations of energy (there cannot be infinite momentum). Then, if there is an upper limit constraining the variation of momentum below some finite value, the perturbative computations (via Feynman diagrams) are always finite.

Interestingly, as Witten argues, the ordinary uncertainty principle when applied to strings gets modified. Indeed, uncertainty in position in the string case becomes[4]:

$$\triangle x = \frac{\hbar}{\triangle p} + \alpha' \frac{\triangle p}{\hbar}.$$

The string introduces the new parameter $\alpha' \frac{\triangle p}{\hbar}$ that fixes some minimal distance at which string interactions occur. This is consistent with what we saw in the previous chapter about the fact that string theory's physical content admits a minimal length in spacetime:

$$x \sim 2\sqrt{\alpha'}.$$

As long as α' is not zero, that is, insofar as we stick to energy scales in which string physics is dominant, the quantum field theory's divergencies arising from zero-distance interactions disappear.[5]

Then, the idea of replacing point particles with strings produces computational tractability. Interestingly this idea also has philosophical consequences insofar as it unveils the fact that quantum string theory delivers a physical ontology more fundamental than that of particles physics. All sort of particles arise from all sort of vibrational modes of underlying strings. As I said, differently from other approaches to quantum gravity, string theory quantizes gravity by also delivering a unifying physical ontology.

Now, string theory is a quantum theory whose laws and equations can be formally obtained from the quantum strings' action pretty much in the same way in which ordinary quantum mechanical laws and equations can be formally obtained from the quantum particles' action. However there is an important difference between the two cases. On the one side, the quantum particles' action is the result of quantization procedures applied to the classical particles' action. The latter is a dynamical structure of a genuine physical theory, i.e., classical Newtonian physics. Within a physical regime in which systems travel at speeds not comparable to the speed of light and in which spacetime distances are of ordinary size, Newtonian laws are empirically robust because their predictions are successfully tested. On the other side, the quantum strings action is the result of quantization procedures applied to the classical strings action. The latter is not a dynamical structure of a genuine physical theory. Indeed, classical string theory is only a toy model without physical content and predictive power, the one that tells a story about fictional entities, namely, classical strings. It is only via quantization of the non-physical classical action that one gains the genuinely physical theory of quantum strings with its laws and its genuine predictive power.

Strings if represented classically are usually described as one-dimensional objects moving in spacetime and tracing out surfaces called worldsheets. The term "worldsheet" is imported from the vocabulary of special relativity, where the term is used to denote the entire history of a physical system during its lifetime. Apparently, within the classical string scenario, spacetime is considered to be a fixed arena somehow prior to string dynamics. This representation of spacetime is similar, *mutatis mutandis*, to that conveyed by ordinary quantum field theory. However, classical string theory does not convey only this representation of spacetime. Indeed, representing strings classically does not necessarily require to embed their worldsheets in spacetime. Classical string scenarios without external spacetime are also possible within classical string theory. As we will see in the next section, insofar as we stick to the fictional case of classical strings, the choice between the two perspectives (the embedded and the non-embedded perspectives) is really a matter of taste. However, as soon as we transit into the physical quantum string scenario (as of Chapter 3), the quantum theory reveals its genuine metaphysical commitment about space and time. Extrapolated from its physical content, that commitment tells that space and time are not a fixed arena, somehow prior to material interactions. And more importantly, if there exist some fundamental "arena" of the physical world, spacetime is not part of it.

2.2 Two philosophical perspectives on classical strings worldsheets

Classical string worldsheets can be represented in two ways, mathematically equivalent but conceptually very different. A possible representation is that of the string worldsheet seen as a surface embedded in spacetime. The two-dimensional character of the worldsheet is due to the fact that the string is a moving one-dimensional object. So, there are two independent parameters describing its evolution. There is a parameter internal to the string, that is, the one moving along the string, also called the "spatial" parameter σ, and there is also a parameter moving transversally to the string, along the direction of the string shift, also called the "temporal" parameter τ. The string moving in spacetime sweeps out a surface. The picture below shows a portion of the surface swept out by an open string (distinct endpoints). The vertical parameter, i.e., the "spatial" parameter σ, varies along the string, whereas the horizontal parameter, i.e., the "temporal" parameter τ, varies transversally to the string[6]:

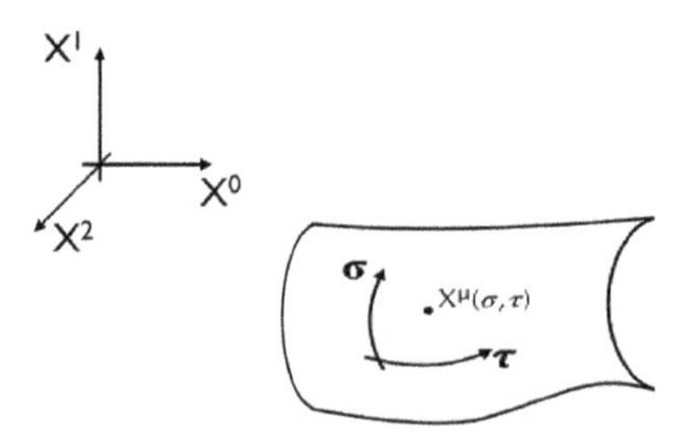

As I said above, the representation is that of an embedded worldsheet, that is, there is an external spacetime arena in which the surface is inserted. In the picture, the external arena appears to be three-dimensional, with a temporal dimension X^0 and two spatial dimensions X^1 and X^2). The finite sheet is the open string's worldsheet swept out during some chosen time interval, say, between $X^0 = t_i$ and $X^0 = t_j$. A canonical way of embedding this worldsheet is that of identifying each of its points (τ, σ) to some numerical value $X^\mu(\tau, \sigma)$ taken by some smooth function $X^\mu(..,..)$ defined on the string's worldsheet. In other words, the function $X^\mu(..,..)$ is the embedding of the string's worldsheet in spacetime because it assigns a spacetime coordinate $X^\mu(\tau, \sigma)$ to any point (τ, σ) over the worldsheet. The embedded worldsheet turns out to be provided with an *induced* metric determining distances among any of its points. The metric is induced because it is inherited from the external spacetime's metric.

However, the very same function $X^\mu(..,..)\text{:}(\tau, \sigma) \rightarrow X^\mu(\tau, \sigma)$ can be read in a quite different way. Still sticking to the classical scenario, a second representation of the worldsheet is possible. In this case, there isn't any posit about any external spacetime arena. The string worldsheet is a non-embedded, two-dimensional manifold. This different formal reading is totally legitimate because what necessarily and sufficiently defines a topological manifold does not include the property of being embedded as submanifold in some higher spacetime.[7]

So, from this second perspective, all we have is the string worldsheet. The $X^\mu(\sigma, \tau)$s values are not spacetime coordinates; rather, they are values of scalar fields filling the non-embedded string worldsheet at any point. Since there is no external arena, the string does not move. Any variation of any value $X^\mu(\sigma, \tau)$ is the variation of some field's value at that point. From this perspective, classical string theory is the same as a two-dimensional classical field theory – see also Witten (1996). One might say that any posit about the existence of a spacetime arena in which strings dynamics unfold turns out to be redundant even in a classical scenario. If away from the string's worldsheet, there is no spacetime, then there is no such a thing as an induced metric on the string worldsheet. "Distances" among any of its points are indeed defined in relation to a structure internal to the worldsheet, also called the *auxiliary* "metric" (Witten, 1996).

As we will see, the internal and induced metrics both show up in the classical action of the string. However, the two interpretations assign to each metric different physical meanings. According to the non-embedded worldsheet interpretation, the "induced metric" is an internal product among fields over the string. What defines a distance over the worldsheet in this case is the internal (auxiliary) metric. By contrast, according to the spacetime embedding interpretation, the induced metric is responsible for the metrical structure of the worldsheet.

In the below sections, some formal basics of the classical bosonic string are presented according to both perspectives. As I said, within the classical non-physical paradigm, neither of them has any representational privilege. However, this equivalence breaks down as we transit to the quantum string. This is due to the fact that the quantization of the classical string action produces a genuine physical theory making contact with phenomena, although indirect. From the physical quantum perspective, the picture of strings physically embedded in some spacetime arena and coupling with background fields somehow prior to them turns out to be a fictional representation of string dynamics. Since quantum string theory is a unifying account of all fundamental interactions, those background fields (including the metric one) have to be contained in the string quantum spectrum of states. This important point will come back in the next chapter.

2.3 What is the action of a physical system?

The action of any physical system is an intrinsic feature of its dynamics. The mathematical nature of an action is similar to that of a functional, that is, loosely speaking, that type of function mapping vectors onto numerical values. Indeed, mathematically, any action associated to a physical system transforms its dynamical history (the input of the functional) into a real number (the output of the functional). Actually, to be precise, the functional applies to the whole set of physically possible dynamical histories of the system, assigning to each history a number. Let's unpack this point.

After the consolidation of the Newtonian description of the laws of nature in terms of instantaneous relationship, physicists noticed at some point that in addition to a system's instantaneous state, its dynamical history could also have been described. This description would be conveyed by a function satisfying mainly two requirements. First, once the system is chosen, this function should always be the one taking initial and final values corresponding to the observed initial and final states of the system. Second, the function should always take (along the path branching out from the initial to the final state) the lowest possible value.

As this new formal language kept settling, the philosophical attitude toward scientific inquiry also transited from analyzing physical properties of systems at some instant in time to understanding those properties in terms of the systems' entire dynamical histories. In the history of physics, one of the first experimental settings in which this emerging methodology has been applied is relative to the dynamical history of a light beam going through two different media. In this case, the function that gets minimized is the time T for the beam to get from A to B. There are an infinite number of possible paths that the light might take. It turns out that light invariably takes the path on which the function T takes its lowest value.

The first set of results about light behavior led physicists to wonder whether this property is a prerogative of light or whether it is a regularity shared by many other types of systems. It turned out the insight was correct. Any system can be described by a function that gets minimized. Nature always chooses the trajectory along which this function is minimized. Now, by definition, this function is the action of a physical system, and the law-like statement about nature's choices is known as the principle of least action. The principle applies to any mechanical system. It applies to classical mechanics, where the action is a function defined by the integral of the difference between the potential and kinetic energies of a system – the potential energy describing the system's interactions and the kinetic energy telling us about the system's state of motion. In this case, the principle turns out to be equivalent to Newton's laws.

The principle also applies to quantum physics where the action is defined by a slightly more complex formal setting. This is because in quantum mechanics, a system does not follow a single path whose action is stationary; rather, its behavior depends on all physically possible paths and on the values the action takes on them. In this case, the action corresponding to the various paths is used to calculate the path integral. The latter is a computational tool giving the probability amplitudes of the various physically possible outcomes (the notion of path integral will be analyzed in more detail at the end of this chapter).

Concluding this section, action in physics is the thing that is minimized: the propagation time for a light beam or the average energy of some body skating across a hilly surface between two different sites. Newton framed the laws of nature around questions like "where was the particle at time t?" or "what was its momentum?", and so on. The subsequent introduction of the notion of physical action changed the way in which laws of nature are framed. Indeed, the action of a physical system frames the physical laws around a function taking values on all the intermediate states of the system as it dynamically evolves from the initial state to the final one.

2.4 Classical action of a free particle

Based on the previous section, we can now take some further step. I start with defining the classical action of a particle for then generalizing this dynamical structure to the string case. For the sake of simplicity, let's assume our classical particle is a free one – which means the particle does not interact with anything and for this reason its action is devoid of potential energy. The particle has mass m and moves in a gravitational field (technically it is not completely free as it couples with a gravitational field, but let's pretend the coupling is neglectable).

The particle moves tracing out its world line in a D-dimensional spacetime. The action describing its motion is the following integral:

$$S = -m \int dt \sqrt{-\frac{dx^{\mu}}{dt}\frac{dx^{\nu}}{dt} g_{\mu\nu}(x)}, \tag{2.1}$$

where the metric tensor $g_{\mu\nu}(x)$ represents the gravitational field. Also, the x^{μ} and x^{ν} are the spacetime coordinates of the evolving position of the moving particle. Finally, t is the particle's *proper time* – namely, the time measured by a clock which is travelling attached to the particle – so the particle is stationary with respect to the clock.[8]

This action has two important formal properties. It is Lorentz invariant, and it is also invariant by reparameterization of the proper time.[9]

By means of the formal identity,

$$dt \sqrt{-\frac{dx^{\mu}}{dt}\frac{dx^{\nu}}{dt} g_{\mu\nu}(x)} = ds,$$

in which ds is an infinitesimal portion of the particle's world line (here the identity is not proved), one can rewrite the action S in the following way:

$$S = -m \int ds.$$

This reformulation of the action reveals the physical relation between the particle's action and its world line: the point particle sweeps out its world line and its dynamics is described by an action proportional to the length of that world line.

However, the action S as it appears in both reformulations has two disadvantages. First, in the formulation (2.1) the presence of a square root makes quantization difficult. Second, in both expressions the action obviously cannot be used to describe a massless particle. So, physicists usually formally manipulate S as it appears in (2.1) by introducing a formal factor that fixes the problem. This factor is an auxiliary temporal metric $\eta(t)$. This modification produces an action which is formally different, namely,

$$S' = \frac{1}{2} \int (\eta^{-1} \dot{x}^{\mu} \dot{x}_{\mu} - \eta m^2) dt, \tag{2.2}$$

where $\dot{x}^{\mu} \equiv \frac{dx^{\mu}}{dt}$.

In spite of the formal differences, the actions S and S' are still physically equivalent. It frequently happens in physics that two actions having different formal structures and different number of degrees of freedom turn out to be physically equivalent, and string theory does not constitute an exception. As we will see, the formal mismatch between two physically equivalent string actions produces consequences for the classical string and also for the quantum one. However, these consequences in the classical string scenario are conceptually different from those analyzed in the quantum one. This is probably due to the different conceptual natures of the two scenarios. On the one side, one has a classical string model producing a non-physical scenario, and on the other a quantum physical theory describing fundamental physical interactions. In the classical string case, the formal mismatch between two physically equivalent actions explains why conformal symmetry is built into classical string theory since the start. This point will be unpacked in Chapter 3.

2.5 Classical action of the bosonic string

The classical string action is given in this section by means of generalization of the point particle case.[10]

Loosely speaking, open strings can be visualized as rubber bands having two distinct end points, whereas closed strings might be seen as circular rubber bands (coinciding endpoints). In other words, if the "spatial" parameter σ is a periodic variable (in the sense that $X^{\mu}(\ldots)$ values have a periodic dependence on σ), the string is closed. If instead the parameter σ varies within a finite, non-periodic interval, the string is open.

Let's assume that a string system evolves within some spacetime arena. Similar to the point particle case, the string's action can be formulated by showing its proportionality to the worldsheet's area. Such formulation is called the Polyakov's action[11]:

$$S_P = \frac{T}{2}\int d\tau d\sigma\sqrt{-\gamma}\gamma^{ab}(\sigma)g_{\mu\nu}\partial_a X^{\mu}\partial_b X^{\nu}, \tag{2.3}$$

where, as we saw, τ and σ are the string worldsheet parameters, T is the string's tension, γ_{ab} is the "metrical" structure internal to the worldsheet – also called auxiliary metric and such that $\gamma = det\gamma_{ab}$ (analogously to the particle case).

Also $g_{\mu\nu}$ is the metric tensor (or gravitational field) of the D-dimensional spacetime in which the worldsheet is embedded. The metric tensor represents any arbitrarily chosen Riemannian spacetime geometry. This arbitrariness of geometry in the classical string action is a symptom of the lack of physical meaning characterizing the model. No role is assigned in the classical setting to the physical geometries arising as solutions to the Einstein field's equations. This point will come back in Chapter 3 within the analysis of general relativistic spacetime emergence from the quantum string action. For now, I only want to emphasize a crucial point about the role played by the Remannian tensor metric in the classical string action. It is an independent degree of freedom because it has not physical meaning. Then, its arbitrariness completely tampers any arguments against background independence of string theory based on the claim that geometry would be put in the theory by hand and that this *a priori* status of geometry would be somehow inherited by the physical quantum action. The methodological construction of the classical toy model of strings does not say anything whatsoever about the genuine physical and metaphysical lesson that string theory provides about the non-fundamentality of space and time. Moreover, the *a priori* role of non-physical geometry is abandoned when moving into the quantum theory. As we will see in Chapter 3, the quantum string perspective geometry is an a posteriori, derived physical structure.

Defining $h_{a,b} = g_{\mu\nu}\partial_a X^\mu \partial_b X^\nu$ as the induced spacetime metric on the worldsheet, one obtains the Nambu-Goto action:

$$S_{NG} = -T \int d\tau d\sigma \sqrt{-det(h_{ab})}. \tag{2.4}$$

Now, although they differ in the number of degrees of freedom, the two actions are physically equivalent. So far, we provisionally assumed that the string worldsheet is embedded in some spacetime arena. In this case, S_P describes the string's propagation in an arbitrarily dimensional spacetime.[12] If read in light of the opposite assumption that the string worldsheet is a non-embedded surface, both string actions instead describe a field theory over the latter, where the $X^\mu s$ represent all the values taken by scalar fields at any point of the surface.

2.6 Symmetries of the classical string action: defining conformal symmetry

Similar to the actions of point particles, both classical string actions S_P and S_{NG} have symmetries. As I said above, symmetries are transformations of either spacetime coordinates or string parameters preserving the physical content of the action. The classical string symmetries mainly include, but not limited to, those of the classical particle action.

Starting with the classical Nambu-Goto action S_{NG}, it turns out that it is invariant with respect to the Poincaré group of transformations. As I specified in the section about particles (in one of the endnotes), this group contains linear combination of Lorentz transformations and translations,[13] i.e.,

$$X'^\mu(\tau,\sigma) = \Lambda^\mu_\nu X^\nu(\tau,\sigma) + a^\mu. \tag{2.5}$$

Now, from the embedded manifold perspective, Λ^μ_ν is simply a D-dimensional Lorentz transformation of spacetime points and a^μ is a spacetime translation. Here, D is the dimension of the external spacetime containing the worldsheet as a submanifold. From the non-embedded perspective instead, these transformations appear to be global symmetries internal to the manifold. Since there is nothing outside the manifold, they simply exchange free fields propagating along the string. This classical symmetry (in both perspectives) will be maintained by the quantization process. That guarantees that quantum particles' states produced by string vibrations are intrinsically characterized by mass and spin.

Another symmetry of S_{NG} is its invariance through diffeomorphic changes of the string parameters σ and τ (or reparameterizations),

$$X^{\mu}(\tau,\sigma) \longrightarrow X^{\mu}(f(\tau), g(\sigma)) = X'^{\mu}(\tau', \sigma'), \tag{2.6}$$

where $\tau' = f(\tau)$ and $\sigma' = g(\sigma)$.

As the Nambu-Goto action, the classical Polyakov S_P is invariant through Poincaré transformations. Both readings above (from the embedded and the non-embedded perspectives) apply to this action as well.[14]

However, differently from the Nambu-Goto action, the classical S_P is also *conformally symmetric*. As we will see in the next chapter, this symmetry is actually built into the classical S_P since the start. The symmetry will not be maintained by the quantization of the classical action. But, as we will see, its reimposition at the quantum level is the sufficient and necessary condition to consolidate the predictive power of the quantum theory. Indeed, its reimposition will crucially contribute to produce general relativistic spacetime emergence in quantum string theory.

For now, I need to extrapolate from standard mathematical literature some basic definitions about conformal transformations. They will be tailored here to the string case. The rationale is that this basic formal terminology will turn out to be quite useful to understand general relativistic spacetime emergence analyzed in Chapter 3, and spacetime emergence through non-commutative string theory presented in Chapter 7.

The first definition to be introduced here is that of *Weyl transformation*. The latter might be defined in terms of a function acting on the internal metrical structure of the string worldsheet in the following way:

$$\gamma_{ab} \longrightarrow e^{\pm 2\omega(\sigma)}\gamma_{ab}. \tag{2.7}$$

This transformation rescales of a factor the metric tensor γ_{ab} of the string worldsheet. To keep notations simple, I am here using σ as denoting (τ, σ).

The rescaling factor can be written in the following general way:

$$\omega(\sigma) = (D-2)\phi(\sigma).$$

The physical meaning of this rescaling factor can be visualized as the action of stretching, or contracting, distances over the string worldsheet. Hence, replacing that factor in the general expression of the Weyl transformation above, we get

$$\gamma_{ab} \longrightarrow e^{(D-2)\phi(\sigma)}\gamma_{ab}.$$

D is a number that plays a role in showing in what sense conformal symmetry is built into the classical string action since the start. In the general case of a sub-manifold embedded in a higher-dimensional space-time, D is the dimension of the sub-manifold on which distances are dilated or contracted. In the string worldsheet case, the dimension D of the sub-manifold is equal to 2. This holds true also in the case in which the string worldsheet is seen as a non-embedded manifold of dimension $D = 2$, where D in this case is not defined in relation to an external higher-dimensional arena, rather to the number of internal degrees of freedom of the field theory over that manifold.

So, the rescaling map over the string worldsheet acts like the identity map. This mathematical fact provides the basic formal explanation of why strings worldsheets are *Weyl invariant.*

Looking at the string worldsheet as a non-embedded manifold, *Weyl invariance* means that however one acts on the internal metrical structure of the worldsheet, either stretching or dilating distances, it does not have any meaningful physical consequence. That is, Weyl invariance is a *gauge invariance* of the internal metric tensor.[15]

Now, *conformal transformations* can be succinctly defined as Weyl transformations in conjunction with diffeomorphic changes of the string parameters (τ, σ). Note well that this characterization seems to entail that conformal transformations form a subgroup of the Weyl ones. But this is not the case since the two sets are different.

A Weyl transformation is a physical change of the metric. It is a transformation that changes the proper distances at each point (τ, σ) by a scalar factor, and the factor may depend on the place (τ, σ) – but not on the direction of the line whose proper distance we measure. So, a Weyl transformation is not a coordinate transformation on the string worldsheet at all. Also, note that a Weyl transformation isn't a symmetry of the usual laws characterizing atomic physics or the Standard model. Indeed, particles are associated with preferred length scale, so particle physics is not scale invariant.

Conformal transformations constitute indeed a subset of coordinate transformations. They include isometries. Isometries are those coordinate transformations $\sigma_i \longrightarrow \sigma_f$ that have the property that the metric tensor expressed as functions of σ_f is the same as the metric tensor expressed as functions of σ_i. Conformal transformations are almost the same thing. But they require something broad, namely, that these two tensors are equal functions up to a Weyl rescaling. That is, a conformal transformation is a map leaving the metric invariant up to scale, hence preserving angles.

A Weyl transformation instead actively scales the metric, keeping manifold internal parameters (or coordinates) fixed. In this sense then, if something is Weyl invariant, along with being invariant by diffeomorphisms, then it is conformally invariant.[16]

2.7 Equations of motion

In this section, I am presenting some formal results that can be stated within both representations of the string worldsheet. Here I derive from the classical action a possible formulation of the bosonic string's dynamics. Conformal invariance is used in this classical context twice – first, to write the equations of motion in a particularly simple form and second, to find an important classical constraint supplementing these equations. However, the mathematical results of this section are only stated. For details concerning their mathematical derivation, I refer the reader to the bibliographic references contained in the endnotes.

Deriving equations of motion of the bosonic classical string requires few simple steps: consider one of the two classical actions seen in the previous section, and take its variation with respect to X^μ. Since conformal symmetry is a main ingredient of this analysis, this procedure will be applied only to the Polyakov action S_P. The result will be a wave equation. Then, depending on whether we are considering an open free string or a closed one, we will pick the appropriate boundary condition.[17]

The variation of S_P with respect to X^μ produces initially the following set of equations:

$$\partial_a(\sqrt{-\gamma}\gamma^{ab}\partial_b X^\mu = \sqrt{-\gamma}\nabla^2 X^\mu = 0. \tag{2.8}$$

The metric tensor γ_{ab} internal to the worldsheet turns out to be described by three independent variables.[18]

Then, a convenient conformal gauge for the worldsheet's metric tensor in which the string equations of motion take a simple form is the following[19]:

$$\gamma_{ab} = \eta_{ab} e^{\phi}, \tag{2.9}$$

where the flat metric η_{ab} internal to the worldsheet appears to be multiplied by a positive function. The latter is the scaling factor preserving angles but not lengths mentioned in the previous section. So, γ_{ab} is a *conformally flat* metric.

Then, in this conformal gauge, the action S_P becomes the action of a free theory and the string equations of motion become the familiar wave equation in dimension two:

$$\left(\frac{\partial^2}{\partial\sigma^2} - \frac{\partial^2}{\partial\tau^2}\right) X^\mu(\tau,\sigma) = 0. \tag{2.10}$$

Now, according to some basic functional calculus facts, any solution of a wave equation of this type can be written as a superposition of some left-moving waves and some right-moving waves.[20] More precisely, the superposition can be written as

$$X^\mu(\tau,\sigma) = X^\mu_L(\tau+\sigma) + X^\mu_R(\tau-\sigma), \tag{2.11}$$

where $X^{\mu}_{L}(\tau+\sigma)$ is the left-moving component and $X^{\mu}_{R}(\tau-\sigma)$ is the right-moving one. Some popular functional analysis theorems about partial differential equations establish that any general solution can be always be written as an expansion of *Fourier modes*. This mathematical fact turns out to be particularly useful in the case of string dynamics. Indeed, the latter amount to be an infinite variety of vibrational modes, that is, an infinite variety of periodic signals. And the Fourier series are the most accurate formal approximations of this type of dynamics. Indeed, they represent periodic signals as infinite sums of sine wave components. A periodic signal is just a signal that repeats its pattern at some period. The feature of periodicity captures well the physical properties of the string vibrational modes.

Now, both components of the wave superposition above can be expressed in function of the Fourier modes. The Fourier modes are denoted by α^{μ}_{n} and $\overline{\alpha}^{\mu}_{n}$, respectively the right-moving Fourier mode and the left-moving one:

$$X^{\mu}_{R}(\tau-\sigma)=\frac{1}{2}x^{\mu}+\alpha'\overline{p}^{\mu}(..)+f(\alpha')\sum_{n\neq 0}\frac{1}{n}\overline{\alpha}^{\mu}_{n}(\ldots) \tag{2.12}$$

$$X^{\mu}_{L}(\tau+\sigma)=\frac{1}{2}x^{\mu}+\alpha' p^{\mu}(..)+f(\alpha')\sum_{n\neq 0}\frac{1}{n}\alpha^{\mu}_{n}(\ldots).$$

For notational simplicity, these two formulas have been simplified by omitting formal details. The term x^{μ} is the coordinate of the center of mass of the string, whereas p^{μ} is the total momentum of the string. The presence of position and momentum in the equations of motion shows that strings do not only vibrate but also move as single units. The string vibrational modes are instead reflected into the linear combinations of Fourier modes α^{μ}_{n} and $\overline{\alpha}^{\mu}_{n}$. Note that the index n of the sum does not take only integral values since these are still classical equations. Moreover, n never shows up equal to zero inside the sum in both identities. That is because the zeroth-order Fourier modes are linked to the total momentum of the string, so they are contained in the second term of each sum:

$$\alpha^{\mu}_{0}=\sqrt{\alpha'}\,p^{\mu}$$

and

$$\overline{\alpha}^{\mu}_{0}=\sqrt{\alpha'}\,\overline{p}^{\mu}.$$

Since we are still within the classical string paradigm, the solutions $X^{\mu}(\tau,\sigma)$ are *real* functions. As I said, their Fourier modes show that strings vibrate and that for this reason classical strings can be represented as infinite sequences of classical harmonic oscillators, each sitting at a point of the string, for any string point. In Chapter 3, the canonical quantization process will change the status of these harmonic oscillators. It will transform them from classical harmonic oscillators to quantum ones, and for this reason, they will be represented by linear operators.

This change of status brings into the story a crucial feature of quantum strings: quantizing the string entails computing the quantum spectrum of states of those linear operators. These spectra, for reasons explained in Chapter 3, contain all the existing elementary particles, along with those so far only theoretical predicted by the Standard model. Each type of particle is produced by a specific oscillation frequency n of the quantum string.

When analyzing the equations of motion of a system, inevitably the study includes the set of their boundary conditions. In general, boundary conditions uniquely pick the class of solutions satisfying some specific requirement. In this book, boundary conditions will show up within the quantum string context as I'll be analyzing the physical and mathematical aspects of T-duality and of the holographic duality (anti-de Sitter (AdS)/conformal field theory (CFT) correspondence). Although in this section we are still in the classical string case, I introduce few basics about boundary conditions preparing the ground for future analysis. This preliminary presentation does not require to specify whether the string is classical or quantum.

How to impose boundary conditions on the set of equations of motion for a string? It depends on whether the string is closed or open. If one needs to pick specific solutions for closed strings, one will impose boundary conditions expressing periodicity. As I said, closed strings are like loops; hence, the "spatial" parameter σ along the string is a periodic one. Then, broadly speaking, the boundary condition in this case would be

$$X^{\mu}(\tau,\sigma) = X^{\mu}(\tau,\sigma+2\pi). \tag{2.13}$$

This condition picks from the general set of solutions the subset of those having integral values of wave number n. Then, the periodicity condition picks a set of solutions like the following:

$$X^{\mu}(\tau,\sigma+2\pi R) = X^{\mu}(\tau,\sigma)+2\pi RW. \tag{2.14}$$

The parameter W above is called the winding number. It is an indicator of how many times a closed string has wound around one or more compact extra dimensions. The peculiar wounding dynamics of a closed string will be described in Chapter 4 as I develop an argument for spacetime emergence by means of T-duality.[21]

If instead one needs to pick specific solutions for an open string, there are two kinds of strings configurations selecting different boundary conditions: open strings can have endpoints free to move, but they can also have fixed endpoints. The first case would be a class of solutions $X^{\mu}(\tau,\sigma)$ with a "spatial" derivative vanishing at the string endpoints. This class would be picked by the boundary condition

$$\frac{\partial X^{\mu}}{\partial \sigma}_{|\sigma=0,\pi} = 0. \tag{2.15}$$

The vanishing of this derivative shows that the $X^{\mu}(\tau, \sigma)$ picks the same value on $\sigma = 0$ and $\sigma = \pi$. The endpoints are indeed free to move as there isn't any constraint on the "temporal" derivative of the solution.

A different case is constituted by an open string with fixed endpoints. The class of solutions in this case would be picked by the boundary condition

$$\frac{\partial X^{\mu}}{\partial \tau}_{|\sigma=0,\pi} = 0. \tag{2.16}$$

Finally, a constraint on the string's equations of motion comes from the conformal invariance of its worldsheet. In the classical string setting, conformal symmetry is built into the Polyakov action S_P since the start. As long as we stick to the classical string, the constraint on the equations is nothing more than a formal derivation. However, conformal symmetry and this related constraint gain physical meaning once revised within the quantum string scenario. As we will see in Chapter 3, they both are crucial to show how quantum string theory admits general relativistic spacetime emergence. Conformal symmetry reimposed on the quantum string worldsheet provides the theory with a genuine physical prediction, although it is a gauge symmetry. Usually, genuine physical predictions of a theory arise when some gauge symmetry is broken. In quantum string theory, when the gauge symmetry in play is the reimposed conformal invariance, the story changes completely – see also Huggett and Vistarini (2015).

Let's derive here the formal constraint from the classical symmetry. As I said, conformal invariance means invariance with respect to stretching or dilating distances on the worldsheet. In other words, the varying of the Polyakov action S_P with respect to the tensor metric γ_{ab} internal to the string worldsheet vanishes:

$$\frac{\delta S_P}{\delta \gamma_{ab}} = 0.$$

Now, the energy momentum tensor for the string worldsheet T_{ab} is defined in function of Polyakov action's variation, that is,

$$T_{ab} = -\frac{2}{T}\frac{1}{\sqrt{-\gamma}}\frac{\delta S_P}{\delta \gamma_{ab}},$$

where the parameter T should not be confused with the energy momentum tensor. T is the string tension.[22] One can see that the conformal symmetry of the classical action S_P produces a vanishing energy momentum tensor of the string worldsheet:

$$T_{ab} = 0.$$

This constraint will come back in the physical scenario of quantum string theory described in Chapter 3.

2.8 Rods and clocks are useless on a quantum string worldsheet

According to Witten, when we deal with the ontology of quantum string theory, we should accept things for what they appear, and what it would appear is simply that the $X^{\mu}(\tau, \sigma)$s are fields living on the strings, whose values depend on the stringy "spatial" and "temporal" parameters[23]:

> So spacetime with its metric determines a two-dimensional field theory. And that two-dimensional field theory is *all* one needs to compute stringy Feynman diagrams.

So, from the perspective of quantum string theory, spacetime with its metric might be replaced with a two-dimensional CFT over a non-embedded string worldsheets. The idea of an external spacetime in which these surfaces are embedded is redundant. Positing an external spacetime arena in which strings evolve does not have any explanatory power. One might also take a further step and read that passage in light of the result I will analyze in the next chapter. It is not only a matter of redundancy but also a matter of uneven metaphysical status. General relativistic spacetime is an emergent structure in virtue of being a low-energy derivation from the restored conformal invariance of the two-dimensional quantum field mentioned by Witten. The quantum string worldsheet appears to be more fundamental than general relativistic spacetime.

Now, one can claim that the quantum string worldsheet is more fundamental than ordinary spacetime, but that does not rule out that the string manifold might still contain some spacetime structure, perhaps the one of a less ordinary kind. If that is the case, the talk about spacetime emergence via the identification of some deeper structure (the quantum string worldsheet) would constitute a much weaker argumentative line in favor of string theory's independence from geometry.

Then, it is worthy to analyze more closely the worldsheet structure. In particular, a more precise identikit of the string parameters τ and σ can clarify what kind of emergence talk we have here. As I said, the two parameters are called " temporal" and "spatial", respectively. Is the terminological choice due to the fact that they are somehow time- and space-like structures? They might produce ordinary spatiotemporal relations after all, except for the fact that these relations are confined to a lower-dimensional arena. Then, should we conclude that these parameters bring geometry back into the fundamental scenario of the theory? Let's call them *sigma*-space and τ-time to avoid confusion with ordinary space and time. Answering these questions requires analyzing the worldsheet's conformal symmetry. More precisely, the Weyl invariance entailed by the symmetry.

Now, string worldsheets have causal structure since they can be divided into σ-space-like, τ-time-like, and light-like paths (Huggett, Vistarini, and Ẃutrich, 2013). This causal structure is preserved by Weyl invariance of the internal metric tensor γ_{ab} because the rescaling factor, namely, the quantity $\exp^{\omega(\sigma)}$ in (2.7), is strictly positive; hence, it does not change the line element's sign.

However, Weyl invariance of γ_{ab} does not preserve the line element's lengths. Indeed, as we saw earlier on, Weyl invariance can be thought as gauge invariance with respect to the metric tensor: either stretching or dilating distances does not have any physical consequence. This fact has profound implications for the σ-space and the τ-time. First, the length assigned by the metric tensor γ_{ab} to any curve has no physical significance, since it can be rescaled to anything one chooses. Second, the "proper τ-time" of a τ-time-like curve, that one would expect to be an invariant quantity associated to the curve, is not preserved through Weyl transformations. Then, the notion of proper τ-time is lost in the theory (see Huggett, Vistarini, and Ẃutrich, 2013, page 12).

That said, it is hard to see how the stringy parameters τ and σ might somehow resemble or retain features distinctive of ordinary spatiotemporal relations. Rather, it looks like Weyl invariance of the worldsheet provides the stringy parameters with features deeply incompatible with ordinary space and time. The former lack of those basic metrical properties peculiar to the latter. And so rods and clocks are useless on the string worldsheet because length and duration lose physical meaning.

2.9 Brief description of some quantization procedures

Classical mechanics can be quantized by using different mathematical methods, all producing the same quantum mechanical theory. A popular method is that of canonical quantization. The formal procedure amounts to be a sort of "promotion" of functionally simple mathematical objects, like numbers, to functionally complex mathematical objects, like linear operators. More precisely, physical properties of a system that in classical mechanics are described by functions taking numerical values, for example, the numerical value of the total linear momentum of a system at a certain time, turn out to be described by linear operators taking vectorial values in quantum mechanics. In other words, quantum physical properties are linear operators mapping states of the system onto other states.

The deep formal difference between the classical and quantum representations of physical properties is grounded on the radical physical difference between classical and quantum behaviors. Many physical aspects of quantum behaviors do not have a classical counterpart, so they require a tailored mathematical language. Loosely speaking, this language is provided by linear algebra over infinite-dimensional vector spaces.

Now, classical strings get quantized via the canonical approach as any other classical system. A classical string can be formally represented as a sequence of classical harmonic oscillators. When the string action is quantized via canonical method, each of them gets promoted to the status of quantum harmonic oscillators (each represented by a linear operator). This method will be presented in basics in Chapter 3 as I will be describing how to obtain the spectrum of all the possible quantum states of a string.[24] The goal there is to show that one of those vibrational states produces quantum geometry.

Another popular approach to quantization of classical mechanics uses instead the formalism of path integrals. The quantization procedure applies to any classical actions of any classical systems.[25] I already introduced the notion of action in Section 2.3, along with the principle of least action. There I said that both notions apply to classical and quantum systems but that in quantum mechanics, their formal uses are slightly different. Quantizing a classical action via path integral procedure is a consequence of the fact that any quantum system does not follow a single path whose action is stationary; rather, its behavior depends on all physically possible paths and on the value of their action. The path integral gives the probability amplitudes of the various physically possible outcomes.

Now, the path integral procedure has been already indirectly described in Section 2.1 as I was explaining why traditional quantization procedures do not apply to gravity. Computing the path integral consists in calculating a series (or sum) of perturbations (of increasing energy) of the classical action.

But what does it mean to perturbate the classical action of a system? In Section 2.5, we saw that the classical dynamical equations of a system (that is the laws obeyed by the system) can be obtained by taking a suitable formal derivative of the system action. This set of equations have physical solutions, namely, all the possible dynamical evolutions the system can follow (each associated to some initial condition) in virtue of its obeyance to the corresponding classical laws. A perturbation of a classical action amounts to be a perturbation of one of its classical solution by means of putting energy in the system – for example, entangling the system with the surrounding environment or changing the configuration of its interactions by increasing them. This produces a higher-energy behavior deviating from the original classical laws, since the perturbed evolution of the system is not a solution of the original laws anymore. In other words, you take an on-shell configuration of the system (satisfying the original classical laws represented by the initial equations of motion), and then by progressively putting an increasing amount of energy in the system (successive perturbations of increasing energy), you produced an off-shell system configuration (not satisfying the original classical laws anymore). The path integral procedure is used in Chapter 3, where I reconstruct the derivation of general relativistic spacetime from the quantum string action, derivation originally presented by Gasperini (2007) and Polchinski (2005). Here, a basic description of this procedure in the string case prepares the ground for future analysis.

The procedure has several steps. First, we start imposing a purely instrumental restriction, that is, a pure mathematical embedding of the string worldsheet in a target flat spacetime. By doing so, string points gain target spacetime coordinates $X^\mu(\sigma,\tau)$ – which are functions of the worldsheet's parameters σ and τ. In this way, the procedure gains two degrees of freedom. It depends on the choice of a specific mathematical embedding and also depends on the choice of the worldsheet "metrical" structure.

Now, broadly speaking, an integral is a summation procedure functionally depending on one or more variables ranging over some interval. A path integral is a summation procedure functionally depending on a variation of possible paths (or possible evolutions) of some physical system. More specifically, the procedure replaces the classical computation of a system's action, the one summing quantities over the system's actual trajectory, with a sum over an infinity of quantum-mechanically possible trajectories. The final goal of the path integral is the computation of a quantum amplitude.

The path integral in the string case is usually expressed in the following way:

$$Z = \int [dg(\sigma,\tau)dX(\sigma)]e^{-S[g,X]}, \tag{2.17}$$

where the integral is taken over all the possible string worldsheet "metrics", here denoted by $g(\sigma,\tau)$, and over every possible embedding X^μ : $(\sigma,\tau) \in \Sigma \hookrightarrow X^\mu(\sigma,\tau) \in X$. Moreover, Σ is the string worldsheet, and X is the target spacetime.

Second, the action S showing up in the exponential's argument of the path integral is the Polyakov action. In this context, the action is provisionally rewritten because of the mentioned instrumental restriction, that is, the purely formal requirement that strings propagate against a flat target spacetime. So, in this way, the internal worldsheet tensor $\gamma_{ab}(\tau,\sigma)$ becomes the Euclidean metric $g_{ab}(\sigma^1,\sigma^2)$, and this turns out to simplify the calculations.

So, the Polyakov action showing up in the exponential's argument is rewritten in the following way:

$$S = \frac{1}{4\pi\alpha'} \int_\Sigma d^2\sigma \sqrt{g} g^{ab} \partial_a X^\mu \partial_b X_\mu, \tag{2.18}$$

where $d^2\sigma$ stands for $d\tau d\sigma$. This integral is taken over all possible $g_{ab}(\sigma^1,\sigma^2)$ and over all possible formal embeddings $X^\mu(\sigma^1,\sigma^2)$.[26]

Notes

1 This presentation of the bosonic string mostly relies on a popular textbook, namely, "String theory and M-Theory" by Becker, Becker, and Swartz (2007, chapters 1–5). The selected parts are presented through steps given mostly without proofs.
2 Fields' quanta obey quantum mechanical laws, i.e., basically the Scrödinger equations. For this reason, quantum field theory is presented in many textbooks as the theoretical framework obtainable via second quantization, i.e., the further quantization procedure applied to fields after quantizing particles' classical action.
3 As we will examine in the next chapter, it turns out that string theory naturally predicts the existence of gravitons. And not only that
4 See Witten (1996, p. 29).
5 Parenthetically, even if Feynman computations in the context of strings do not produce infinities because of the energy upper limit, still energy scales at which string theory physical content is dominant are much higher than those produced by our presently available experimental technologies.
6 The parameter τ is called "temporal" because in many embedded representations, the direction transversal to the string is usually the temporal one, with respect to the external time.
7 Contemporary topology considers manifolds as devoid of any ambient embedding. Historically speaking, this is mainly due to the fact that mathematicians like Gauss and Riemann discovered at some point that there are intrinsic properties of manifolds that can be studied independently of any particular embedding.
8 See Becker, Becker, and Swartz (2007, page 18).
9 These are fundamental symmetries in particle physics. Loosely speaking, they are those changes of coordinates preserving the physical content of the particles models. To be precise, this action is not just Lorentz invariant but Poincaré invariant. A Poincaré transformation is something more general than a Lorentz transformation. It is indeed a linear combination of a Lorentz transformation and a translation. For this reason, Poincaré transformations are also called inhomogeneous Lorentz transformations. This group of symmetry characterizing particle physics will be appropriately extended to the string case.
10 What follows as far as the end of this section is based on Becker, Becker, and Swartz (2007, pages 19–30).
11 $S_P \sim \int ds^2$; that is, the action minimizes area. In the S_P expression above, the indexes a and b refer to the string worldsheet, whereas the indexes μ and ν refer to the spacetime arena.
12 Within the classical regime, there is no dimensional constraint. As we will see, what sets the dimension of the bosonic theory equal to 26 ($D = 10$ in the case of superstring) is the quantization process.
13 What follows about symmetries of S_{NG} is based on Becker, Becker, and Swartz (2007, pages 24–25).
14 What follows about symmetries of S_P is based on Becker, Becker, and Swartz (2007, pages 30–31).
15 When $D > 2$, we can define an *infinitesimal* Weyl transformation, which simply is an infinitesimal rescaling of the manifold metric tensor. The latter is obtained by differentiating with respect to ω, $\delta\omega : \gamma_{ab} \longrightarrow (D-2)e^{(D-2)\delta\omega}\gamma_{ab}$. These infinitesimal transformations will be used in Chapter 3, among other things, to derive a class of general relativity solutions from quantum string theory.

16 In preparation of Chapter 3, it might be useful to explicit the rationale behind the choice of privileging conformal transformations in this section. As we will see, reimposing conformal symmetry on the quantized string's action is crucial to proving that the theory admits emergent general relativistic spacetime. Indeed, reimposing that invariance produces an important constraint on the kind of classical gravitational fields that can couple with the classical string within the fictional classical representation of strings dynamics. As I said, this fact bypasses a quite common objection against string theory background independence, namely, that through the perturbative formulation of the quantum theory, geometry is put in the classical action by hand at the beginning of the quantization process.

17 The content of this section is a reformulation based on Becker, Becker, and Swartz (2007, pages 31–34).

18 The worldsheet has two internal independent variables, and the tensor has two indexes. Now, since the tensor is a symmetric 2×2 matrix, it should be described in principle by four independent variables. However the action chosen is conformally symmetric. Conformal symmetry means invariance of the physics with respect to stretching and contracting the internal metric tensor γ_{ab}. Then, the number of degrees of freedom of the internal metric tensor decreases from four to three. For a more detailed analysis of this point, see Becker, Becker, and Swartz (2007, page 31), in particular equation (2.23).

19 See Becker, Becker, and Swartz (2007, page 59, equation (3.2)).

20 In dimension one, any general solution looks like $g(t,s) = g_L(s - vt) + g_R(s + vt)$, where $g(,)$ is a function of space and time, s is the spatial coordinate, and v is the velocity.

21 As I said, string theory posits the existence of compact extra dimensions added to the four-dimensional general relativistic spacetime. But remember that four-dimensional spacetime and compact extra dimensions share the same emergent physical nature. Positing something as real and claiming that it is not fundamental are not contradictory claims. As we will see, the methodology used to show general relativistic spacetime emergence is different from the strategy showing the emergence of the compact extra dimensions. In the second case, string dualities are crucial. Now, particularly in the context of T-duality, the set of boundary conditions for closed string dynamics mentioned above are part of the formal argument in favor of the compact extra dimensions emergence.

22 See Becker, Becker, and Swartz (2007, pages 26–36).

23 "Reflections on the Fate of Spacetime", Witten, 1996, page 27.

24 See Becker, Becker, and Swartz (2007, section 2.4). The method uses equal time commutators and constraint equations to get the Hilbert space of the physical states.

25 The following schematic presentation of this procedure is based on "Quantum Field Theory," by Srednicki, Cambridge University Press, 2007.

26 This class of metrics has signature (++). The advantage of using a Euclidean path integral consists in having an integral which is well defined almost everywhere. In fact, the Euclidean metric can be not singular even in cases of topologically non-trivial surfaces.

3 General relativistic spacetime emergence

"So we arrive at a quite beautiful paradigm. Whereas in ordinary physics one talks about spacetime and classical fields it may contain, in string theory one talks about an auxiliary two-dimensional field theory that encodes the information. The paradigm has a quite beautiful extension: a spacetime that obeys its classical field equations corresponds to a two-dimensional field theory that is conformally invariant. If one computes the conditions needed for conformal invariance of the quantum theory derived from the Lagrangian, assuming the fields to be slowly varying on the stringy scale, one gets generally covariant equations that are simply the Einstein equations plus corrections of order α'" (Witten, 1996, page 28).

3.1 Sketch of the chapter

In this chapter, I argue that quantum string theory admits emergent general relativistic spacetime. The argumentative line chosen will show that we don't need to restrict our attention to the non-perturbative formulation of string theory (via the holographic duality) in order to find features of background independence in the theory. Indeed, as I argue in this chapter, the theory's perturbative formulation shows features of background independence as well, although they are different by those identified via dualities arguments.

Now, the object to which perturbation applies is any arbitrary Remannian classical geometry which is put in the classical string action by hand. The classical string action is a non-physical model, so the fact that its classical metric tensor is an arbitrary free parameter neither constitutes a counterexample against the background independence of the quantum string action nor it is a feature inherited by the latter: physical geometry is not put in the quantum action by hand since the start.

Emergence in this chapter develops around the derivation of the geometrodynamical structure of general relativity from the restored conformal symmetry of the string quantum action – the derivation considered here is the one presented by Polchinski (2005) and by Gasperini (2007). As I said in the previous chapter, the moral of the story behind the restoration of conformal symmetry is different from that one might expect. Usually, it is the breaking of some gauge symmetry that produces physical consequences. Here instead conformal anomaly of the quantum string does not account for experimental facts, whereas restoring conformal symmetry does. Conformal symmetry reimposition has profound physical consequences in quantum string theory since it produces general relativistic spacetime emergence. In this sense, this gauge symmetry challenges the general idea that gauge symmetries are simply formal features of the way in which a theory's physical content is formally represented – see also Huggett and Vistarini (2015).

The line of thoughts unfolding in this chapter starts with presenting how quantum string theory predicts the existence of gravitons. In other words, the theory predicts the existence of quantum geometry. Then I show, by unravelling Polchinski and Gasperini works, how the quantum string dynamics dictate that the gravitons' collective behavior at large distance is governed by the Einstein field equations (EFEs). These two properties combined together constitute the formal structure and physical content of general relativistic spacetime emergence.[1]

My philosophical analysis of this derivation identifies a notion of emergence combining two features. On the one side, it is mainly a form of supervenience-based emergence; on the other, the novelty appearing at the higher level is identified in epistemological terms. To this second reason, the analysis of the inter-theoretical relation between general relativity and quantum string theory plays an important role. The existing philosophical and scientific literature on emergence on which I rely in this chapter to unpack the issue is mainly constituted by three different, quite popular works. They don't deal with string theory; still each offers philosophical insights on emergence smoothly combining with my identification criteria of general relativistic spacetime emergence in string theory.

The first two studies are by Robert Batterman, in his "The Devil in the Details. Asymptotic reasoning in explanation, reduction and emergence" – see Batterman (2002) – and by Jeremy Butterfield, in his "Less is different: Emergence and Reduction reconciled", along with "Emergence, Reduction and Supervenience: a Varied Landscape" – see Butterfield (2011a, 2011b). Although being different notions of epistemological emergence, the two characterizations proposed by the authors share the idea – which is also central to my analysis – that a theory can be considered to be emergent as long as it appears to be definable as limit of some underlying, finer theory, typically as some parameter in the latter goes to some crucial value. According to Batterman, in order to speak of emergence the limit needs to be singular, whereas that is not required by Butterfield. The third approach considered here is that presented by Gordon Kane in "Supersymmetry. Unveiling the ultimate laws of nature" – see Kane (2013). His method of effective theories can be seen as a sort of epistemological counterpart of my form of metaphysical foundationalism, concisely expressed in Chapter 1 by the claim that reality comes into layers.

3.2 General relativistic spacetime emergence – part I: how quantum string theory predicts the graviton

Sometimes in theoretical physics, the formal articulation of a theory on one side reveals the deepest features of its physical content, and on the other, it relies on metaphysical assumptions about reality. In these cases, the theory's mathematical language turns out to be richer than a purely logical one and also more rigorous than ordinary language. Indeed, it is the main vehicle of physical information often not directly accessible by means of measurements and also counterintuitive. This is particularly true for string theory.

The ontologically primitive entities of the theory (in the sense of not being analytically and physically derived from anything else in the theory) are quantum strings. Strings are shown by the theory's physical content as more fundamental than spacetime. The latter reappears as derived structure emerging from their underlying dynamics. String theory ontology also includes higher-dimensional objects, called D-branes, but the latter are not ontologically primitive in the theory – at least according to my reading. The existence of D-branes and the main schools of thought about their status in the theory are analyzed in Chapter 5 within the context of emergence through holographic duality.[2]

Here, I don't consider their dynamical contribution for simplicity reasons, and this choice does not compromise the completeness of my argument in favor of general relativistic spacetime emergence. Indeed, this emergence deeply involves the nature of gravity in quantum string theory, and gravity is produced by closed strings. The latter cannot attach to D-branes, that is, their spectra do not contain any modes associated with D-branes fluctuations and dynamical behavior.

It is natural to wonder at this point where string dynamics unfold. Let's say for now that wherever they take place, if traditional geometrical notions like "where" and "when" still make sense, this arena, if any, is not general relativistic spacetime because according to the physical content of the theory, general relativistic spacetime arises from quantum string dynamics. As I said in the previous chapter, a key strategy in string theory bypassing divergency problems is that of replacing elementary particles with strings. More precisely, particles arise as excitations of strings, or strings' oscillation modes "produce" particles. Perhaps better, particular configurations of vibrating strings define string models of particles which in a low-energy regime reproduce the chiral fermions and gauge bosons of the Standard model. In particular, inside the second group, a particle with zero mass and spin 2 arises as oscillation mode of closed strings.

Now, the particle that would carry the gravitational force would have these mass and spin features. This theoretical fact has been known by physicists for a long time. This theorized particle is called the graviton. So, in this sense, string theory *predicts* the existence of the graviton.[3] The graviton is identified with a particular closed string state inside the lowest energy level. Here, I am accounting for this identification in a simplified way and by following Zwiebach (2009, ch. 6) and Susskind's lectures – "String theory and M-theory".[4]

Now, the identification between the graviton and one of the particles showing up at the lowest energy state of a closed string, relies on the traditional method used to identify photons as the lowest energy excitation state of an open string. That is because in physics, gravitons are usually represented by pairs of entangled photons – we will see in basics how that works later on in this section. Therefore, by using simplified models, we first see the photons case, then the graviton one.

Photons are relativistic particles; hence, we would need to describe relativistic open string dynamics. But the description can be hugely simplified as long as we stick to a fictional scenario. One may use the fictional representation of a relativistic open string mathematically embedded in a three-dimensional space, where x, y, z label the three spatial dimensions. The string oscillates in three dimensions and also shifts along the z axis. The fictional character of this model lies in the fact that the string is spatially embedded and there is no time dimension.[5]

In this simplified model, we immediately face the need of a further simplification. Quantizing directly relativistic oscillations of a string would be too difficult. We need to find an indirect strategy that produces the same result, that is, some familiar setting in which non-relativistic oscillations are described by familiar periodic functions, and then quantize these oscillations. Here, I shall not introduce explicitly these functions. It suffices to say that quantizing non-relativistic string oscillations basically amounts to promote some numerical coefficients of those functions to the status of linear operators (annihilation and creation operators). This method is the first quantization procedure I sketched in Chapter 2, i.e., the canonical approach (see also the Appendix for more detail).

Now, in order to get a non-relativistic description of relativistic strings oscillations, all we need to do is to boost the string along the z axis with infinite momentum.[6] This fact along with a further property peculiar to strings produces a situation in which the relativistic string can have a bi-dimensional non-relativistic representation of its oscillations.[7]

The string's peculiarity is the fact that there are no independent degrees of freedom concerning oscillations along the z axis, that is, in this settings they are all perpendicular to z. In other words, in this model of representation the non-relativistic two-dimensional motion over the x, y plane exhaustively encodes all the physical properties of the string's oscillations in three dimensions. So, eventually we end up with the simplest possible model of a string's oscillations, one that is two dimensional and non-relativistic.

Now, let's recall that in Chapter 2, we saw that every solution to the string's equations of motion can be expanded in Fourier series. The Fourier modes α_k^μ showing up in those series are functions encoding the strings vibrations. Quantizing string dynamics according to the canonical approach is done by promoting these functions to linear operators, namely, creation and annihilation operators. In this two-dimensional, non-relativistic model, these linear operators describing quantized oscillations are mathematically decomposed in two components, one along the x axis and the other along the y axis. So, the creation and annihilation operators act along both axes (see also the Appendix). Precisely, in this setting, we have the creation operator a_n^+ along the x axis and the creation operator b_n^+ along the y axis. Also, we have the annihilation operator a_n^- along the x axis and the annihilation operator b_n^- along the y axis – here n is simply the oscillation's frequency.

Now, let's focus on the creation operators a_n^+ and b_n^+. They create two excited states of the string sharing the same energy value n. An energy string state is mathematically represented by a vector. So, the two operators above acting on such vector will produce two energy vectors (called energy eigenstates), both relative to the same energy value (see the Appendix for more detail). Since we are looking for pairs of entangled photons, we want to find the lowest energy states of the open string. So, we feed the ground state $|0>$ of such a string into the cheapest creation operator, namely, a_1^+ and b_1^+. Once they are fed, these operators add into the ground state just one unit of energy. And that's all we need for our purpose. The state we get is

$$a_1^+|0> .$$

This state has vectorial properties along the x direction and the lowest value of energy. But this is not the only state with the lowest energy level. We also have

$$b_1^+|0> .$$

This state has the same energy as the one above, but it has vectorial properties along the y direction.

These two vectors are two linearly independent energy eigenvectors relative to the same energy level. That entails that if we linearly superpose them, you get another vector with the same energy, i.e., you get

$$(a_1^+ \pm b_1^+)|0> .$$

This new state represents a string which has the property of a linearly polarized particle.

Moreover, taking a linear superposition of $a_1^+|0>$ and $b_1^+|0>$ is not the only way in which we can get a string state at the same energy level. Indeed, it is still possible to combine them in a circular way, i.e.,

$$(a_1^+ \pm ib_1^+)|0> .$$

This state represents a string which has the property of a circularly polarized particle. In both cases (linear and circular polarizations), the polarization is transverse to the direction of motion of the string. Note that there are only *two* superposition states we can produce without changing the energy state. Therefore, the particle polarized in those two ways must be massless, since it has two spin states (i.e., +1 and −1). Indeed, a massive spin 1 particle would also have a third spin state, namely, the spin-0 state. So, what we got shows we have a massless particle.

Moreover, $a_1^+|0>$ and $b_1^+|0>$ are two perpendicular vectors (one along x-axis, the other along y-axis), both transversal to the direction of the boost of the string, i.e., the z-axis. Finally, they can both rigidly rotate about the z axis. So, following a simplified line of reasoning, there is only one conclusion consistent with Lorentz invariance, with relativity, and with the properties of $a_1^+|0>$ and $b_1^+|0>$: the states $(a_1^+ + b_1^+)|0>$ and $(a_1^+ + ib_1^+)|0>$ behave like the polarization states of a photon.[8] So, they can be identified with photon-like objects.

Now, the usual and simplest way of representing the graviton in physics is the following. One takes two photons, either both circulating in the left-handed sense (i.e., with total spin equal to −2) or both circulating in the right-handed sense (i.e., with total spin equal to 2). Then, one sticks them together and lets them move along some axis. Then one has a graviton. In other words, a graviton moving along some axis with maximum angular momentum is mathematically the same as two entangled photons, each of them having one unit of angular momentum. And a couple of equally handed photons mathematically appear in the lowest energy state of a closed string. Computing a closed string's spectrum of states is pretty much the same as computing that of an open string. However, part of the formalism relative to open strings slightly changes when considering closed strings. Indeed, the functions describing oscillations double for the latter: we have a right-moving oscillator (i.e., moving along the increasing direction of the string's parameter) and a left-moving oscillator (i.e., moving along the opposite direction). Then, for each frequency n of oscillation, there are two creation operators, namely, a_{-n}^+ (connected to the left-moving oscillator) and a_n^+ (connected instead to the right-moving oscillator). The same apply to the annihilation operator, i.e., we get a_{-n}^- and a_n^-.

The lowest level of excitation of a closed string contains the following particles:

$$(a_1^+ + ib_1^+)(a_{-1}^+ + ib_{-1}^+)|0>,$$

and

$$(a_1^+ - ib_1^+)(a_{-1}^+ - ib_{-1}^+)|0> .$$

These are two circular superpositions of photons: the first made by two right-handed circularly polarized photons (adding to angular momentum 2), and the second made by two left-handed circularly polarized photons (adding to angular momentum –2). Both superpositions mathematically represent the graviton.

Parenthetically, in the same energy level, there are also two zero-spin particles, both still some superpositions of entangled photons. One is called the dilaton, namely, the famous hypothetical particle first introduced by Kaluza-Klein theory, and then predicted by string theory. This particle is the quantum of the background field dilaton Φ assumed by string phenomenology. The second particle is called the axion, and it is the hypothetical quantum of the antisymmetric field $B_{\mu\nu}$, a further background field also living in the string phenomenology. The dilaton field Φ will show up again as crucial component of a physical scenario explored in Section 3.3.1. There the rationale behind this exploration will be that of sharpening the precise conceptual role of conformal symmetry in quantum string theory.

Concluding this section, quantum string theory predicts the existence of the graviton. This prediction is one of the two structural components of general relativistic spacetime emergence in string theory. The fact that string theory predicts gravitons means that quantum geometry can really be found as dynamical byproduct of quantum strings. In other words, geometry can be explained by quantum string theory, rather than just being posited. The next section contains the second component of general relativistic spacetime emergence.

3.3 General relativistic spacetime emergence – part II

The second structural component of general relativistic spacetime emergence is the derivation of general relativity's solutions from the lowest energy regime in the perturbative structure of the quantum string action. My analysis of this formal derivation heavily relies on two important works: "String theory, Superstring theory and beyond", vol. I (see Polchinski, 2005, pages 108–118) and "Elements of String Cosmology" (see Gasperini (2007, Chapter 3).

This section articulates in three parts. Section 3.3.1 is an inquiry about the status of conformal symmetry in quantum string theory. Reimposing conformal symmetry of the quantum string action is the final step of the quantization process. Is this step necessary? Can we have quantum string theory in the presence of conformal anomaly? Answering this question requires a brief exploration of some less popular and more exotic string theory scenario. The analysis of this topic relies on Huggett and Vistarini (2015). Here, the review of the paper's outcomes will be connected to general relativistic spacetime emergence.

Section 3.3.2 presents a simplified version of the formal derivation of the EFEs, namely, the derivation of the general relativistic laws. Indeed, I'll only focus on deriving the vacuum solutions from the reimposed conformal symmetry of the quantum bosonic string. General relativistic spacetime can indeed be obtained from the quantum string action in the presence of matter fields as well. This more complex (and more realistic) derivation mimic the same logic as the simplest case. So, for simplicity reasons, here I will stick to the simplest case.[9] Quite importantly, this derivation will reveal that the underlying quantum string theory poses an unambiguous constraint on what kind of classical gravitational field must dominate the low-energy physical scenario emerging from the quantum string action, namely, Einstein gravitational field.

Now, Sections 3.3.1 and 3.3.2 combined together will support the concluding claim that not only gravitons are found in the quantum string spectrum, but also that quantum string theory (its perturbative formulation) dictates what low-energy laws must be obeyed by their low-energy collective behavior, namely, the EFEs. Finally, in Section 3.3.3, I interpret philosophically in what sense one can apply the notion of emergence in this context.

3.3.1 *Reimposing conformal symmetry: necessary or optional?*

Let's recall from Chapter 2 that conformal symmetry amounts to be the Weyl invariance of the worldsheet "metric" tensor combined with the diffeomorphism invariance of the two string parameters. Conformal symmetry is built into the classical string action since the start. However, the quantization of the classical action breaks classical Weyl invariance, hence breaking conformal symmetry too. Restoring conformal symmetry *via* reimposing Weyl symmetry at the quantum level it produces the derivation of the EFEs.

In this section, I want to explain the rationale behind the choice of restoring conformal symmetry. Do we restore conformal symmetry of the quantum action as an independent postulate useful to derive the gravitational field result? Or, instead, is conformal symmetry an intrinsic property of the quantum string, something without which there would not be a quantum string theory at all?

A positive answer to the first question produces a scenario in which general relativistic spacetime is emergent in virtue of its irreducibility to quantum string theory. Indeed, in this case, a third principle (conformal symmetry) independent from the underlying quantum string theory would contribute to produce the derivation. Quantum string theory can only partially explain general relativistic spacetime. This kind of emergence might fit a supervenience-based scheme connoted by some robust ontological character in virtue of the partial reducibility to the basal quantum string dynamics.

A positive answer to the second question, instead, yields a different idea of emergence. Emergent general relativistic spacetime is reducible to quantum string theory since the restored conformal symmetry is a constitutive part of the theory. In this case, the supervenience-based scheme is still applicable, but the physical novelty introduced by the emergent general relativity laws can only be identified in epistemological terms, namely, by appealing to the fact that it does not have any counterpart in the underlying scenario produced by the quantum strings laws. The novelty is a symptom of the incommensurability of the two physical scenarios, the supervenient and the basal ones. This incommensurability is an inevitable consequence of the physical change of energy scale needed to transit from the quantum string dynamics to those of Einstein gravitational field.[10] In this section, I argue for a positive answer to the second question. In other words, I argue that reimposing conformal symmetry on the quantum string action is a necessary condition – see also Huggett and Vistarini (2015).

In chapter 2, I introduced the classical Polyakov's action S_P, namely,[11]

$$S_P = \frac{1}{4\pi\alpha'} \int_\Sigma d^2\sigma \sqrt{-\gamma}\gamma^{ab} \partial_a X^\mu(\sigma) \partial_b X^\nu(\sigma) g_{\mu\nu}(X). \tag{3.1}$$

Based on the previous section, we know that a complete expression should also include the antisymmetric field $B_{\mu\nu}$ and the dilaton field Φ, i.e., the other two fields introduced by the classical string phenomenology (in virtue of the fact that the quantum string dynamics physically predict their quanta). But for now, they are both set to zero for simplicity reasons.

As we saw, this action requires an auxiliary "metric" γ^{ab} internal to the string worldsheet Σ. For this reason, the action S_P has a number of degrees of freedom which are bigger than those of the Nambu-Goto action, i.e., the other classical string action introduced in Chapter 2. As I said there, the two classical actions produce two equivalent scenarios. For this reason, it must be the case that the additional variables of the S_P action due to the auxiliary metric γ_{ab} cannot be genuine degrees of freedom. So, the S_P action (in addition to diffeomorphic invariance over the worldsheet) cannot depend on the choice of the auxiliary metric γ_{ab}.

But this independence from the choice of the metric means that S_P is the Weyl invariant. That explains why conformal symmetry is built into the classical theory since the start.

The quantization procedure performed by perturbing the classical string action breaks conformal symmetry by breaking Weyl symmetry. In other words, the quantum string observables end up with being dependent on the degrees of freedom introduced by the auxiliary metric γ_{ab}. At this point, by reimposing the Weyl invariance of the quantum action, we rescue the quantum theory's physical content from this dependence. And by doing that, we end up getting the Einstein field result.

But why do we need to rescue the quantum theory from this dependence? To understand the kind of necessity involved here, I show what would happen if we decide to preserve Weyl anomaly during quantization. What follows in this section heavily relies on Polchinski (2015, sections 3.4, 3.7, and 9.9) and Huggett and Vistarini (2015).

One may start by rewriting the classical action above still setting the antisymmetric field $B_{\mu\nu}$ to zero, but this time without setting to zero the dilaton field Φ:

$$S = \frac{1}{4\pi\alpha'} \int_{\Sigma} d^2\sigma \sqrt{-\gamma}[\gamma^{ab} \partial_a X^{\mu}(\sigma) \partial_b X^{\nu}(\sigma) g_{\mu\nu}(X) + \alpha' R\Phi(X). \qquad (3.2)$$

As we saw in Chapter 2, this classical action can be read as a two-dimensional interacting field theory. Now, the more the string is excited, the heavier its vibrational modes are. Since such massive particles have never been met in experiments so far, physicists usually restrict their attention to the lightest modes of the two-dimensional theory. This practical strategy produces some computational benefits, that is, the lowest string vibrational modes can be easily approximated in this case by the model of an effective string action over a D-dimensional target spacetime. This effective action is actually physically equivalent to the worldsheet action truncated to its lowest energy limit. Such effective action looks like the following:

$$S_X = \frac{1}{2} \int d^D X \sqrt{-G} e^{-2\Phi} \left[-\frac{2(D-26)}{3\alpha'} + R + 4\partial_{\mu}\Phi \partial^{\mu}\Phi + O(\alpha') \right].^{12} \qquad (3.3)$$

The presence of a non-trivial dilaton background Φ is crucial to the exploration of the consequences entailed by Weyl anomaly. Indeed, a non-trivial background dilaton inevitably shows up if the condition of Weyl symmetry is released. This fact is due to the connection existing between the trace of the stress energy tensor over the worldsheet and the beta functions of the background fields over spacetime:

$$T^a_a(\sigma) = -\frac{1}{\alpha'} \left(\beta^G_{\mu\nu} \gamma^{ab} + i\beta^B_{\mu\nu} \right) \partial_a X^{\mu} \partial_b X^{\nu} - \frac{1}{2} \beta^{\Phi} R.$$

As we saw in Chapter 2, Weyl symmetry over the worldsheet means that the stress energy tensor of the worldsheet is traceless, namely, $T^a_a(\sigma) = 0$. For that to happen, all the beta functions on the right-hand side have to vanish: $\beta^{G}_{\mu\nu} = \beta^{B}_{\mu\nu} = \beta^{\Phi} = 0$. So, by releasing the constraint of Weyl symmetry over the worldsheet, one is actually releasing the constraint of spacetime background fields having vanishing beta functions. In particular, the non-vanishing beta function β^{Φ} relative to the non-trivial dilaton yields a departure of string theory from the critical dimension ($D = 26$ for the bosonic string, $D = 10$ for the superstring case).

This new theoretical context (also called *non-critical* string theory) can have some inconsistency issues, unless Weyl symmetry is reimposed in more general terms by controlling the divergencies of the dilaton. Indeed, getting a non-trivial, well-behaving dilaton requires Weyl symmetry, but it does not require being back to the critical dimension. Weyl symmetry is restored in arbitrary dimension, and in this new setting, the number of extra dimensions becomes the dynamical outcome of the dilaton's dynamics – see Huggett and Vistarini (2015) for detail.

Let's unravel this point. Without Weyl symmetry, the parameter ω determining the overall scaling becomes another field of the worldsheet coordinates $\omega(\sigma)$ in addition to $X^{\mu}(\sigma)$. The background field dilaton Φ turns out to depend linearly on $\omega(\sigma)$. So, the rescaling factor $\omega(\sigma)$ increases the target spacetime's dimension. But this is not the only thing it does: $\omega(\sigma)$ cannot be actually seen just as one more spacetime dimension in addition to $X^{\mu}(\sigma)$, since there is no translation symmetry along it. Indeed, such symmetry would be entailed by Weyl symmetry, and we just assumed provisionally that Weyl symmetry is not holding here.[13]

This fact is also reflected by the effective action in the new higher-dimensional target spacetime: in the theory defined by this action, the laws of physics appear to be not invariant under translations along the additional dimension $\omega(\sigma)$. Given this anomalous physical feature, the new target spacetime seems to move away from any vaguely realistic representation of low-energy, phenomenical spacetime.

But in addition to concerns about the fact that this model does not make any contact with phenomena, there is also a concern about the internal consistency of the model. As I said, the dilaton Φ depends linearly on $\omega(\sigma)$. This fact yields a divergent behavior of the former, which in turn produces a breakdown of the perturbative structure of quantum string theory.

How does the perturbative structure relate to Φ's behavior? By comparing the actions (3.2) and (3.3), it is possible to see that the string coupling in (3.3) is deeply related to the dilaton by the fact that

$$\alpha' \sim e^{2\Phi(X)}.$$

Therefore, where the dilaton diverges (because of Weyl anomaly) is where perturbation theory breaks down. This is how Weyl anomaly poses a thread of internal inconsistency to quantum string theory.

Trying to control the dilaton divergencies, still keeping the Weyl anomaly, amounts to choose a linear dilaton in combination with an appropriate tachyon profile suppressing the divergent behavior of the dilaton. But the suitable tachyon profile turns out to be solution of some tachyon equations of motion which crucially reveal to be equivalent to the condition for Weyl invariance of the dilaton. In other words, the attempt of keeping Weyl symmetry out of the story inevitably produces its reintroduction.[14]

In conclusion, string theory exists away from the critical dimension provided that an appropriate configuration of background fields is chosen. But fixing the dilaton-tachyon configuration is equivalent to reimposing Weyl symmetry in arbitrary dimension. Then quantum string theory does not seem to exist without Weyl symmetry. Therefore, conformal symmetry is an intrinsic principle of the theory. This fact crucially determines the philosophical features of general relativistic spacetime emergence in string theory.

3.3.2 *Getting EFEs' solutions*

Let's start the derivation of the field result by coming back to the classical bosonic action (3.1). As I said, here I am restricted to the bosonic case for simplicity reasons, although I will sketch the derivation in the presence of fermionic fields – for detail on this, see Huggett and Vistarini (2015). A crucial feature of the classical string action (3.1) is the fact that the classical metric tensor $g_{\mu\nu}(X)$ represents any arbitrary Remannian metric tensor over the target spacetime. That is, according to classical string theory, there isn't any special role played by the Einstein gravitational field. Indeed, the classical action is conformally symmetric for any arbitrarily given metric tensor $g_{\mu\nu}(X)$ of the target space. What follows about quantization of (3.1) sticks to Gasperini's presentation of the path integral method (see Gasperini, 2007, Chapter 3). Let's denote with X^μ the classical solution. The latter is then perturbed by quantum fluctuations of power increasingly high. That is, classical X^μ starts to interact with quantum fields z^μ which are not solutions of the classical action (3.1):

$$X^\mu \longrightarrow X^\mu + z^\mu.$$

Given the dependence of $\partial_a X^\mu$ and of the metric $g_{\mu\nu}(X)$ on X^μ, the perturbations of the latter produce perturbations of the former, namely,

$$\partial_a X^\mu \longrightarrow \partial_a X^\mu + \partial_a z^\mu - \partial_a X^\mu R_{\mu\nu\alpha\beta}(X) z^\alpha z^\beta + \ldots$$

and

$$g_{\mu\nu}(X) \longrightarrow g_{\mu\nu}(X) - R_{\mu\nu\alpha\beta}(X) z^\alpha z^\beta + \ldots$$

The perturbed field derivatives and the perturbed metric tensor are both expressed above in a particular reference system, called by Gasperini the system of Riemann "normal" coordinates. The rationale behind this choice is dictated by mathematical simplicity reasons. Indeed, the formal expression of the expansion in this reference system becomes particularly simple – for more information on this point, see Gasperini (2007, page 76).

The above quantum fluctuations around the classical field produce a new interacting quantum field theory in two dimensions, and hence a new action that can be written in the following way:

$$\overline{S} = S + S_{quadr},$$

where S is the unperturbed action (3.1), and the quadratic part is equal to

$$S_{quadr} = \frac{1}{4\pi\alpha'}\int_{\Sigma} d^2\sigma\sqrt{-\gamma}\gamma^{ab}[\partial_a z^{\mu}\partial_b z^{\nu} g_{\mu\nu}(X) - \partial_a X^{\mu}\partial_b X^{\nu} R_{\mu\nu\alpha\beta}(X) z^{\alpha} z^{\beta}].$$

This quadratic part of the quantum model is responsible for divergencies.[15]

However, the quadratic action can be renormalized adding the appropriate counterterm killing the divergent part, namely, the S_{quadr} can be rewritten as

$$S_{quadr} = S_R + S_{\infty},$$

where S_R is the quadratic action renormalized and S_{∞} is the counterterm action, i.e., the term that absorbs the infinity when the momentum k is too high,

$$S_{\infty} = -\frac{1}{4\pi\alpha'}\int_{\Sigma} d^2\sigma\sqrt{-\gamma}\gamma^{ab}\partial_a X^{\mu}\partial_b X^{\nu} R_{\mu\nu\alpha\beta}(X) \times \left[\lim_{\sigma\to\sigma'} g^{\alpha\beta}_{\alpha'}\int \frac{d^2k}{(2\pi)^2}\frac{e^{-ik(\sigma-\sigma')}}{k^2}\right].$$

Not surprisingly, it is possible to see that this limit is introducing a cutoff in space. And this is exactly what renormalization does: the divergent behavior can be ignored because it takes place below the new established threshold.

Nevertheless, such strategy for eliminating divergencies yields a theory showing dependence on some length scale in space, hence a theory which is not conformally symmetric because not Weyl invariant anymore. It is possible to check explicitly how S_{∞} fails to be Weyl invariant. But in this specific case, the failure can be fixed just in case a *certain crucial condition* is held.

Let's describe this condition precisely.
Since the limit is not computable in dimension $D = 2$, we need to rewrite S_{∞} in dimension $D_{\epsilon} = 2 + \epsilon$.

Following Gasperini (2007, pages 77–78), the limit above in dimension D_{ϵ} becomes equal to

$$\lim_{\epsilon\to 0}\frac{\alpha' g^{\alpha\beta}}{\epsilon},$$

and so

$$S_{\infty}^{2+\epsilon} = -\frac{1}{4\pi\alpha'}\int d^{2+\epsilon}\sigma\sqrt{-\gamma}\gamma^{ab}\partial_a X^\mu \partial_b X^\nu \alpha' R_{\mu\nu\alpha\beta}g^{\alpha\beta}\left[\lim_{\epsilon\to 0}\frac{1}{\epsilon}\right].$$

Since $R_{\mu\nu\alpha\beta}g^{\alpha\beta} = R_{\mu\nu}$ along with the fact that the limit depends just on the variable ϵ, which is not a variable of the integration, the above expression can be rewritten bringing the limit out of the integral as

$$S_{\infty}^{2+\epsilon} = -\frac{1}{4\pi}\lim_{\epsilon\to 0}\int d^{2+\epsilon}\sigma\sqrt{-\gamma}\gamma^{ab}\partial_a X^\mu \partial_b X^\nu R_{\mu\nu}\frac{1}{\epsilon}.$$

Now considering the infinitesimal Weyl transformation defined in Chapter 2 and taking the variation of $S_{\infty}^{2+\epsilon}$ with respect to it, we get

$$\frac{\partial S_{\infty}^{2+\epsilon}}{\partial\omega} = -\frac{1}{4\pi}\lim_{\epsilon\to 0}\epsilon\int d^{2+\epsilon}\sqrt{-\gamma}\gamma^{ab}\partial_a X^\mu \partial_b X^\nu R_{\mu\nu}\frac{1}{\epsilon}.$$

Note well the presence of an extra factor ϵ in front of the integral. This extra factor is there because of the way in which the infinitesimal Weyl transformation works on the metrical part $\sqrt{-\gamma}\gamma^{ab}$. According to the formula above relative to $\delta\omega$, this transformation maps $\sqrt{-\gamma}\gamma^{ab}$ into $(D-2)\sqrt{-\gamma}\gamma^{ab}$. But our dimension D is now equal to $2+\epsilon$; hence, the metric part is mapped as

$$\sqrt{-\gamma}\gamma^{ab} \longrightarrow \epsilon\sqrt{-\gamma}\gamma^{ab}.$$

Now, the extra factor ϵ appearing in front of the integral cancels out with the one in the denominator inside the integral. So, we have

$$\partial S_{\infty}^{2+\epsilon} = \lim_{\epsilon\to 0} -\frac{1}{4\pi}\int d^{2+\epsilon}\sqrt{-\gamma}\gamma^{ab}\partial_a X^\mu \partial_b X^\nu R_{\mu\nu}\partial\omega.$$

Coming back to dimension $D=2$ by setting ϵ identically equal to 0, we get

$$\partial S_{\infty} = -\frac{1}{4\pi}\int d^2\sqrt{-\gamma}\gamma^{ab}\partial_a X^\mu \partial_b X^\nu R_{\mu\nu}\partial\omega.$$

Now, one might grasp in which sense S_∞ is not Weyl invariant (hence non-conformally symmetric) unless a certain condition is imposed. Indeed, to have Weyl invariance, we need to set ∂S_∞ equal to zero.

Note well that we can set the action's variation ∂S_∞ to zero only in one case, namely, when

$$R_{\mu\nu} = 0. \tag{3.4}$$

Now, this important identity above is actually the condition of *Ricci flatness*, namely, the equation representing the vacuum solutions of the EFEs. Therefore, the conformal invariance of quantum string theory can be preserved just in case within the classical, low energy limit, the classical gravitational field is the Einstein gravitational one.[16]

Let's summarize what we got so far. By quantizing the closed string, we found in its spectrum of states a spin 2 particle that can be reasonably identified with the graviton. This is the first important result that shows that quantum geometry can really be found in quantum string theory as dynamical byproduct.

Since quantum string theory is a unifying account of all fundamental interactions, the idea of a gravitational field residing outside the string source and interacting with the string turns out to be fictional when tested against the fundamental string scenario. Indeed, from the underlying, finer quantum string perspective, a classical gravitational field is a coherent superposition of infinitely many gravitons produced by certain closed string oscillation modes.

Now, the derivation of general relativistic spacetime from the quantum string action is the second important result. The quantum geometry found in the theory in form of dynamical byproduct of closed strings vibrations is actually not an arbitrary one. Indeed, it can only be a quantum dynamical structure that at low energy becomes general relativistic spacetime.

Now, the two results combined together constitute the formal structure and physical content of general relativistic spacetime emergence in quantum string theory. What we need to understand now is what type of philosophical notion of emergence can fit this structure.

3.3.3 Emergent general relativistic spacetime: some philosophical remarks

In the first subsection, I identified the kind of necessity underlying the reimposition of conformal symmetry on the quantum string action: far from being an independent postulate added to the formal articulation of the theory, conformal symmetry is a necessary component of that articulation.

Although being a gauge symmetry, conformal invariance of quantum string theory turns out to be much more than simply a feature of the way in which the theory represents its physical content. Indeed, to ensure internal consistency and physical significance of the quantum string dynamics (in critical or in arbitrary dimension), we must reimpose conformal invariance of the quantum action in the first place (with dimensional constraint or more generally without). So, as I said, maintaining conformal anomaly does not have any physical consequence, whereas reimposing conformal symmetry does. And this is the opposite of what one might expect from a gauge symmetry.

Now, in this section, I want to complete the philosophical identikit of general relativistic spacetime emergence. As I said earlier on, general relativistic spacetime emergence appears to fit a notion of supervenience-based emergence, where the novelty is identified in purely epistemological terms. The emergent structure arising as low energy limit of underlying quantum string dynamics brings into play novel physical content which is not straightforwardly reducible to the quantum basal one. And this is mainly due to the fact that the low energy limit built into the derivation introduces an asymmetry between the physical contents of the two theoretical levels.

Now, Batterman and Butterfield, already mentioned in Chapter 1, articulate two general views both fitting my picture above. Indeed, they both use a notion of inter-theoretical relation characterized in terms of approximation among theories. The general idea they share with my approach describes the emergent status of a theory in terms of being the limit of some other underlying theory, as some physical parameter in the latter goes to some crucial value. However, the two authors differ with respect to what about that limit can count as emergence.

On the one hand, Batterman claims that we have emergence only in case the limit is a singular one. In this sense, the transition from the underlying theory to the emergent one would present some formal discontinuities expressing a precise notion of irreducibility. Indeed, emergence in this case happens slightly before the limit, because the limit diverges. The emergent physical properties would appear in the asymptotic domain as the more fundamental explanatory theory approaches the emergent one. So, they are irreducible to the more fundamental theory. The asymptotic domain is a kind of "no man's land" (see Batterman, 2002, page 11), which requires an additional new explanatory theory of the emergent phenomena.

His example of inter-theoretical relation between statistical mechanics and thermodynamics can make this point even more clear (see Batterman, 2002, Chapter 4). The general reductive schema in this case is that of taking the continuous limit of statistical mechanics for the number of particles becoming infinitely large. This limit produces classical thermodynamics which describes the collective behavior of this large number of particles, namely, the physical behavior of a fluid. The reduction shows that any density fluctuation arising from intramolecular interactions occurs only at a very small microscopic scale.

But, as Batterman argues, this is not what happens when considering fluid dynamics around its critical point. Around the latter, fluctuations occur at all scales up to the macroscopic one. So, the critical state is a singularity of thermodynamics at which its smooth reduction to statistical mechanics breaks down. Taking the limit of statistical mechanics cannot produce smoothly those macroscopic new states of matter. We need a third theory, a modified thermodynamics allowing for fluctuations, which explains the dynamics of those new macroscopic phenomena.

So, what we have in this case is the occurrence of emergent properties which need novel explanatory stories. These stories require a third novel theory, the "no man's land," which is asymptotic with respect to statistical mechanics, namely, it is irreducible to the latter. Emergence is then seen by Batterman as a robust physical pattern incompatible with any attempt of reduction. One can have emergence only if reduction abruptly fails.

On the other hand, Butterfield develops a different account. As far as I understand, he seems to argue that physically meaningful emergence should be individuated at a different stage of the limit, namely, after the limit is taken. In this sense, emergence and reduction are compatible. Singularity of the limit does not play any role in defining emergence. Here, "reduction" is interpreted as "deduction" in the Nagel sense. The theory deduced after the limit can be thought as emergent because it describes novel behaviors and regularities which are in some sense unexpected or not captured by the vocabulary of the reducing theory.[17]

Now, similar to Butterfield, I argue that the more fundamental theory explains the emergent one because it unveils the basal dynamics underlying the emergent properties. However, I also claim that this characterization straightforwardly applies not in any case of inter-theoretical relations. In cases like general relativistic spacetime emergence, the explanatory power of the underlying quantum theory turns out to be weaker than that described by Butterfield. Indeed, here the formal methodologies used as bridge between the two theories do not preserve the smoothness of the transition even with the lack of formal divergencies that cannot be fixed. Indeed, one has two sets of laws – the basal and the emergent ones – whose radical difference in terms of energy regimes and physical parameters produces a loss of physical content as the transition occurs. So, the more fundamental theory turns out to be the reducing in a weaker sense than that described by Butterfield: the emergent properties are novelties not only because the fundamental theory vocabulary cannot capture them but also because they do not have counterparts in the basal physical scenario. The incomparability of the two theories' vocabularies provides the epistemological criterion to identify those robustly emergent properties. In other words, there are cases like the one studied in this chapter in which the explanatory smoothness of the original Nagel's deduction cannot be preserved whether or not there are formal divergencies during the transition (via energy limits) from one theory to the other. The lack of explanatory smoothness does not require a singularity à la Batterman.

A third approach presented by Gordon Kane about inter-theoretical relations is called the method of effective theories – see Kane (2013, Chapter 3). The author does not explicitly mention any notion of emergence, but his use of the notion of unpredictability "*in practice*," combined with the idea of explainable "*in principle*," might be read in light of the discussion on emergence reconstructed here. Indeed, Kane's view shares with those considered here the centrality of physical approximation methods in the characterization of inter-theoretical relations.

Indeed, the basic formal criterion to identify inter-theoretical relations in the method of effective theories amounts to be the variation of physical length scales in space. The variation occurs within a spectrum. Each effective theory describes phenomena at a certain length scale in space, and it is insensitive to phenomena occurring at different scales. According to Kane, Dirac equations describing electrons and nuclei can explain *in principle* any chemical process occurring at larger-distance scales. However, those equations cannot *in practice* calculate such higher-level processes because such computations are too complicated to be performed.[18] So, some larger-scale phenomena are not predictable in practice from laws governing much lower-distance scales. When considering an effective theory at a given distance scale, some regularities are found in its domain of application which are new with respect to the domain of application of the underlying theory at smaller spatial scale. Still this theory provides an explanation of why things appear as they appear at a larger-distance scale since it explains where the free input parameters used in the larger-scale predictions come from.[19]

As far as I understand, Kane's method of effective theories might be read as the epistemological counterpart of the form of metaphysical foundationalism I propose in Chapter 1, succinctly expressed by the central claim that reality comes into layers. Also, the idea that each effective theory is insensitive to what happens at different scales, combined with the other idea that, at any point of the spectrum, a basal theory cannot compute in practice higher-level physical properties (although in principle it might explain them), might produce some grist for my view that emergence produces a weaker form of explanation not fitting the smooth Nagel scheme.

As I said, general relativistic spacetime emergence in string theory combines two structural components. On the one side, at the quantum string length scale, gravitons are dynamical byproducts of quantum string dynamics. On the other, at ordinary large-distance scales, gravitons' collective behavior (their coherent superposition) obeys general relativistic laws. Based on these features, I first claim that the philosophical nature of general relativistic spacetime emergence is well captured by a notion of supervenience-based emergence. Let's dig into this point to refine the analysis.

The quantization of the classical string action is performed by quantum corrections of the classical solution. Quantum string theory is the full perturbative expansion (i.e., the expansion which includes all the quantum corrections at each order). The truncation in which we worked by using renormalization and infinitesimal conformal maps is the lowest energy piece of the full expansion (see the second subsection). Note well that this low-energy regime is not yet a classical limit.[20] As we saw, the low-energy truncation of the full quantum string theory turns out to contain an ill-defined quantum action because of the divergences. This fact seems to characterize the truncation as an asymptotic domain à la Batterman in which emergence means failure of the reductive limit.

Now, the divergences of the quantum string action can be eliminated by a procedure of renormalization. And, more importantly, those divergencies are not produced by taking the lower energy (still quantum) limit from the full perturbative expansion to the first-order truncation. Rather, our divergencies are due to the quantization of the classical string action. Then, one concludes that divergencies in this context are not the same divergencies used by Batterman: Batterman's divergencies are non-renormalizable singularities of the reduced theory (classical thermodynamics) at which the reductive scheme to the reducing theory (statistical mechanics) breaks down. In our case, the renormalizable divergencies obtained by quantization of the classical string action are not singularities of the reduced theory, namely, they are not singularities of general relativity.

Once we have a non-divergent quantum string theory, we face the fact that conformal invariance has been broken. By reimposing conformal symmetry and by taking at this point the classical low energy limit, we gain general relativity dynamics. So, one might conclude that the all procedure looks similar to the reduction à la Butterfield. The original singular limit has been smoothed and now, after the limit, we perform the *deduction* of classical general relativity solutions from the reducing quantum string theory. After all, conformal symmetry is an intrinsic principle of quantum string theory, so no third principle independent of the two theories is involved in this derivation. However, although there are similarities with Butterfield's approach, my study case produces a notion of emergence more robust, something coupling with a weaker notion of explanation. There might not be the need for a third principle to explain the emergent properties, but there is no clear way of accounting smoothly for the emergent physical properties because of the total lack of counterparts in the fundamental scenario: physical spacetime disappears down there.[21]

3.4 Conclusion: background independence of the perturbative formulation of the theory

In this chapter, I argue that string theory admits emergent general relativistic spacetime. I argue that the derivation of general relativity from quantum string theory can be philosophically labelled as a type of supervenience-based emergence. Moreover, general relativistic spacetime emergence shows one way in which quantum string theory is background independent, namely, its perturbative formulation does not posit any fundamental geometry. This point needs to be unpacked since it actually contains two different background independence claims.

The first claim is that the perturbative formulation of quantum string theory definitely shows *background field independence*. Indeed, as we saw, the quanta of any classical background fields (including those string-inspired) are found at the quantum string scale as dynamical byproducts of string vibrations – and among them, more importantly for our purpose, the graviton. Moreover, we found out that the quantum string dynamical law imposes some explicit constraints on the macroscopic collective behavior of the gravitons, namely, they must collectively evolve according to general relativity dynamics.

The second claim is that the perturbative formulation of quantum string theory shows *background independence*, namely, independence from geometry also in the absence of fields. Indeed, the fact that the theory does not posit any fundamental flat space geometry appears quite clearly in the derivation of general relativity solutions. First, the flat Minkowski geometry belongs to the class of Ricci flat solutions derived from the quantum string dynamics and also imposed by this dynamics as a constraint on the emergent manifest image of the world. In other worlds, flat space geometry is a byproduct of quantum string dynamics. Second, the most common objection against string theory's background independence mistakenly interprets a purely instrumental feature of the classical string action (which is a purely fictional model) as revealing aspects of the fundamental physical ontology of the quantum theory. And so, quite illogically, the fact that the non-physical classical action contains any arbitrary Remannian geometry for purely instrumental reasons is used to argue that geometry is put in the quantum physical theory by hand. I already argued in Chapter 2 that this is not the case. To reiterate the point here, the arbitrary Remannian geometry in the classical action is a free parameter put by hand in virtue of the non-physical nature of that action. Therefore, the concluding claim of this chapter is that the perturbative formulation of quantum string theory is background independent in the strongest possible sense.

3.5 Appendix: what empirical prediction means in this context

The notion of prediction frequently mentioned in this chapter is explained here in more basic terms. Strings can be in many different states of oscillation (or vibration) depending on how much energy they gain from interacting with other strings. The more energy the string has, the faster it oscillates. We can mathematically compute the entire spectrum of oscillation states of a string. In this sense, this is the string energy spectrum. The particles appearing in each of these states composing the string energy spectrum are the particles predicted by the theory.

Now, some of them are actual physical particles already observed in nature, some others have never been observed so far. However, their hypothetical existence is not only predicted by string theory, but it is also postulated by different quantum theories of particles.

Similar to quantum mechanics, a quantum string system is identified to an infinite dimensional vector space. However, a particular quantum state of the quantum string system is not just identified to a vector in that infinite-dimensional vector space, rather to a sub-vector space of the same. That is because a quantum string state can be characterized by infinite modes of oscillations. Let's dig into this.

How does an actual physical particle formally appear in a string quantum state? It appears in the formal guise of a vector of that sub-vector space representing the quantum state of the string. That vector shows some specific mathematical properties universally recognized as representing some key physical property of the actual physical particle in question. Usually, the key physical property of the particle represented by the vector's formal features is its own physical spin. And this holds true also for hypothetical particles, like the gravitons, which have been theorized to exist with certain physical features involving their spin.

Now, quantum string theory borrows part of its mathematical language from quantum field theory. I said in Chapters 2 and 3 many times that the string "produces," or "creates," a certain kind of particle by vibrating in a certain way. That explains why in the chapter when representing mathematically a particle as a string state we used the particle-creation and particle-annihilation operators of quantum field theory. Indeed, in quantum string theory, the mathematical appearance of a particle is nothing else but a particular combination of creation operators acting over the string ground state.[22]

How do creation and annihilation operators formally connect strings and particles? One may visualize a string (either open or closed) as a finite continuous line of points. The line is continuous because it is the result of a continuous limit of points. The limit is taken by making the distances between any two contiguous points infinitely close to zero. Then, one may imagine that sitting on top of each point there is some harmonic oscillator. So, a string can be seen as a continuous infinite collection of harmonic oscillators, each of them oscillating with infinite modes of oscillation (or frequencies). The kind of particles produced by the string will depend on the oscillation frequency of its harmonic oscillators. The string can oscillate with different frequencies at each different point. To reiterate a point made above, that explains why a quantum string state gets identified to a sub-vector space, rather than just to a vector.

As I said in Section 3.2, since we want to compute the spectrum of states of a quantum string system, the harmonic oscillators must be promoted from the status of simple functions (taking numerical values) to the status of linear operators taking vectorial values, i.e., having vectors as outputs. In this sense, the string oscillations produce particles. Remember, particles are formally found in the string energy spectrum in the formal guise of vectors. Then, as quantum field theory teaches, any linear operator can be expressed as combination of annihilation and creation operators. So, this is how the oscillation modes of each point on the string – which classically are just coefficients – become combinations of creation and annihilation operators. And there is a creation and annihilation operator for each frequency of oscillation of the string at each of its points.

Notes

1. The analysis of this formal derivation is done here only in the simplest case, namely, the formal derivation of the vacuum solutions of the EFEs.
2. D-branes are dynamical objects in their own right, and they are important in string theory for several reasons. The following is not an exhaustive list, but it can give a broad idea of why they are important. First, the ends of fundamental strings can attach to them. It has been showed that Yang-Mills type theories (like electromagnetism, weak and strong interactions) involve open strings that are attached to D-branes. Second, they carry the basic charges of a special class of fields which string theory necessarily describes. Finally, in a regime of strong string coupling, D-branes can become lighter than the string itself, and so their behavior can dominate some features of the low-energy physics.
3. Some basic detail on what "prediction" means in this context is contained in the Appendix of this chapter.
4. http://www.youtube.com/watchv=25haxRuZQUk, Lecture 2.
5. Many string theory textbooks use the same instrumental strategy. They introduce string dynamics as physically embedded in some spacetime arena, whose structure is defined by some metric tensor. According to this fictional representation, the metric tensor is a background field somehow prior to string dynamics. In a quantum perspective, this scenario turns out to be unambiguously false. Since quantum string theory delivers a unifying model of all fundamental interactions, every background field interacting with the string, included the metric tensor, must be contained in the spectrum of states associated with the string oscillations. This is the only truthful perspective we should not forget as we are working within the simplified, fictional case. Identifying gravitons in the string spectrum within this fictional scenario does not entail that gravitons are really propagating in spacetime.
6. In this way, we are representing dynamics in the infinite momentum frame or light cone reference frame.
7. More precisely, the projection onto the plane x, y of the string motion is not relativistic because, roughly speaking, the momentum along z is so much higher than the momentum in the other two directions that the only relativistic effects worthy of notice are those along z.
8. These properties are here just sketched. Loosely speaking, $a_1^{+}|0>$ and $b_1^{+}|0>$ look like the x and y components of an electric field produced by a particle moving along z.
9. For an extensive analysis of the more realistic case involving matter fields, see Huggett and Vistarini (2015).
10. Chapter 6 will dig even more into this notions of supervenience-based emergence, irreducibility, and incommensurability.

11 This action is also called the string sigma-model. In the expression above, the coefficient in front of the integral appears slightly different from that in Chapter 1, since it has been rewritten by using the identity $\alpha' = \frac{1}{2\pi T}$.

12 By varying this spacetime action with respect to its arguments it is possible to verify the physical equivalence with the low energy worldsheet action: see Polchinski (2005, section 3.7). As I said, $B_{\mu\nu}(X)$ is set to zero.

13 Translation symmetry is entailed because multiplicative scaling of e^{ω} is the same as an additive shift to the coordinate $\omega(\sigma)$.

14 For detail, see Polchinski (2005, page 323) and also Huggett and Vistarini (2015, section 4).

15 So, we can say that the divergence of the quantum action comes from the one-loop diagram. Let's briefly see in this endnote how the divergence shows up. Let's think about the diagram in position space. The propagator for a scalar particle is

$$\langle z^{\alpha}(\sigma) z^{\beta}(\sigma') \rangle \sim_{\sigma \to \sigma'} log|\sigma - \sigma'|.$$

If the scalar field runs in the loop, the starting and ending points are the same. So, the propagator diverges as $\sigma \to \sigma'$.

16 The same EFE derivation can be performed in the presence of matter fields – see Callan et al. (1985), Huggett and Vistarini (2015). Starting with both gravitational and fermionic fields, we end up with having a heterotic string theory. In such a theory, not just conformal invariance holds true, but also supersymmetry because of fermions. Again, by imposing the requirement of conformal invariance will produce at the first order of α' a condition including the stress energy tensor of the Yang-Mill field, i.e.,

$$R_{\mu\nu} + \frac{\alpha'}{2} tr(F_{\mu\nu}^2).$$

So, conformal invariance entails that at the first order of α', we get EFEs.

17 "I shall take emergence to mean: properties or behaviour of a system which are *novel* and *robust* relative to some appropriate comparison class. Here 'novel' means something like: 'not definable from the comparison class' [...]" (see Butterfield, 2011a, page 3).

18 See Kane (2013, page 44).

19 For example, a tentative theory of atoms predicts the size of an atom by using the value of a parameter, the electron's mass, which is given to it by the underlying subatomic theory. In this sense, the atomic theory gives a description of how things work at the atomic scale, but it cannot explain why they work as they do because it cannot explain why the electron has the mass in question. An explanation of that can be only provided by the underlying sub-atomic scale description that successfully predicts the value of that mass.

20 Parenthetically, note well that focusing on the (non-classical) low-energy regime of the full perturbative expansion does not conflict with the methodology used in the previous subsection. Although there we moved in the other way around, that is, from the classical theory to its quantum formulation, the lowest energy regime (still quantum) in which we settled is the same.

21 The Nagel deduction scheme might not be smoothly applicable to the explanatory scheme proposed in my study case also because the centrality of physical limits in the emergence process produces a notion of derivation only partially reflected by the purely logical features of the Nagel deduction.

22 Note that the ground state in string theory is not exactly the vacuum. It should be thought as the state of an unexcited oscillator. Also because of the

uncertainty principle in conjunction with some quantum fluctuations, it is not entirely correct saying there are no oscillations. The ground state of a string is characterized by very tiny vibrations; hence, you can find some small amount of energy or mass there. However, the lowest energy level of a string remains the one you get applying to the ground state the creation operator which adds into the latter one unit of energy.

4 T-duality and emergence

In the previous chapter, I argued that quantum string theory admits general relativistic spacetime emergence. In this chapter, I argue that quantum string theory admits a complementary notion of spacetime emergence. Here, I argue for the emergent nature of the extra dimensions added by the theory to general relativistic spacetime. The main structure of the argument is grounded on the existence of string dualities. In this chapter, I only consider T-duality.

4.1 Emergent extra dimensions: some preliminary remarks

Extra dimensions play a crucial role in quantum string theory. Postulated mainly for consistency reasons, they turn out to reinforce the theory's predictive power. Indeed, they crucially contribute to let the theory display the massless spectrum of gauge bosons and chiral fermions of the Standard model.

What does it mean to say that extra dimensions are added to the four-dimensional general relativistic spacetime? The basic formal structure of this addition amounts to be a simple Cartesian product, something like $S \times K$, where S is the four-dimensional general relativistic spacetime, and K represents the compact extra dimensions.

Leaving aside for a moment their property of compactness, the mathematical relation between the stringy high-dimensional spacetime and the general relativistic one (also called ordinary) is formally described by a map of projection onto the first factor, namely,

$$\begin{array}{c} S \times K \\ \downarrow^{l} \\ S \end{array}$$

This formal relation represents a low energy physical limit. The limit produces the manifest image of the world, that is, the world appearances obtained by means of measurements and dynamical generalizations.

Now, the geometrical structure S appears in the manifest image, but the extra dimensions K do not. Their formal disappearance (from the scenario of macroscopic laws) is due to the fact that the extra dimensions are not detectable at low energy. As we will see, their property of compactness encodes their physical property of extreme smallness, i.e., of being detectable around the Planck scale.

Then, the moral of the story here is that the high-dimensional stringy spacetime is more fundamental than the ordinary one in virtue of its not being manifest, that is, of its being a higher-energy structure from which the manifest one is derivable by means of a low-energy limit of the higher-energy dynamics. However, although the extra dimensions do not belong to the manifest image of the world, they are emergent as well according to quantum string theory. The stringy spacetime $S \times K$ neither is proved nor is posited by the theory to be at the fundamental level. Indeed, the extra dimensions are emergent. This thesis of emergence will be proven here by T-duality.

4.2 Poincaré's underdetermination problem

As I said in Chapter 1, there is an old story about a possible mismatch between the manifest image of the world produced by the dynamics and its actual structure. The story is known as the Poincaré underdetermination problem. In the original Poincaré formulation, the underdetermination applies to the amount of information we can get about the correct geometry of the world by looking at empirical data, that is, the original problem is about epistemology of geometry. My use of the Poincaré problem in the context of string dualities is at right angle with its original formulation. Indeed, the lesson I want to draw in this book is about physics.

In Chapter 1, my interpretation of the Poincaré problem is applied to the case of a general Hamiltonian. The logic underlying that train of thoughts is the same as the one used here in the more specific context of some strings' Hamiltonian and some strings' duality. In this chapter, I analyze the scenario of underdetermination arising from T-duality. Two string theories showing two inequivalent geometrical arenas, $S \times K_1$ and $S \times K_2$, that is, two theories introducing stringy spacetimes having geometrically inequivalent extra dimensions, may turn out to be experimentally and physically indistinguishable. This possibility is also expressed by saying that the two string theories, although containing two spacetime arenas geometrically inequivalent, are *dual*, where duality is a relation of physical equivalence whose basic formal structure will be discussed in Section 4.3. As we will see, this fact says a lot about the emergent nature of those spacetime arenas.

Let's pretend for a moment that these two dual theories are the only competitors for being the fundamental description of the world. Also let's pretend that each theory posits its own geometry as fundamental, and that we are making an inquiry about which one of these geometries is the correct one – for the sake of the argument, we are also assuming that there is such a thing like a correct fundamental geometry of the world.

This case would be an instantiation of Poincaré's parable since – given their dual relation – any dynamical generalization based on the shared physical content would not reveal what posit about geometry is the correct one. The duality between the two theories tells that string dynamics do not detect geometrical differences between the two sets of extra dimensions. In other words, geometry is underdetermined by strings physics.

And yet lower-energy physical properties produced by vibrating strings, like masses, charges and so on, also depend on extra dimensions. So, one may argue that it is not the geometrical structure of the extra dimensions that contributes to produce phenomena. Although this context is slightly different from that presented in Chapter 1 (here we focus on the extra dimensions), still the underlying logic is the same: even if we posit that the geometry of some set K of extra dimensions is fundamental – hence pretending that the higher-dimensional arena $S \times K$ is fundamental in the theory – still that geometry would not account for the manifest physical phenomena (low-energy physics).

Note well that the point is not simply that string theory does not give facts about the correct fundamental geometry of the world. The lesson is more radical than that: string theory does not give facts about whether there is any fundamental geometry at all. The fundamental structure of the world in the string scenario may well be not geometrical at all. As I said in Chapter 1, a much weaker structure, like a topological one, seems to be in some cases all the theory needs to account for phenomena. But in some other more radical scenario of background independence, not even topology seems to play any explanatory role.

Before unravelling this central issue, I would like to reconstruct briefly the original Poincaré problem, as I think that can enlighten my re-interpretation of it.[1]

For the sake of accuracy, I should mention that the first broad formulation of the underdetermination problem is due to Duhem, a French physicist contemporaneous to Poincaré.[2] Duhem argued that the problem of scientific underdetermination poses serious challenges to our efforts to confirm theories in physics. As Quine suggested later on, this challenge applies not only to the confirmation of physical theories, but also to any claim expressing scientific knowledge.[3]

Differently from Duhem, Poincaré formulated an underdetermination problem within the back then prominent debate about epistemology of geometry.[4] His proof in favor of consistency of non-Euclidean geometries appeared on the scene of the debate right before the advent of general relativity. The main goal of his work was that of undermining all those claims of logical inconsistency of non-Euclidean geometries. In particular, one of the most popular critiques was the one that Poincaire developed against the legacy of Kant, namely, the view that Euclidean geometry is the correct geometrical structure of the world.

What did Kant mean when arguing that Euclidean geometry is the correct geometry of the world? It is widely known that one of the main goals that Kant pursued in the First Critique was that of unearthing the a priori foundations of Newtonian physics. The latter describes the physical fundamental structure of the world in terms of Euclidean geometry.

How did Kant achieve the a priori foundations of Newtonian physics? As far as I know, Kant argued that our understanding of the physical world grounds not merely on experience, but on experience combined with "a priori" concepts. Here, I will not analyze in detail his transcendental arguments. I will just mention that he argues that the possibility of sensory experience depends on certain necessary conditions – which he calls "a priori" forms. He also argues that these a priori conditions structure our experiences of the world. The former allow the latter to hold true.[5]

As he claims in the "Transcendental Aesthetic," space and time are not derived from experience but rather are its preconditions.[6] Experience provides those things which we sense. It is our mind, though, that processes this information about the world and gives it order, allowing us to experience it. Our mind supplies the conditions of space and time to experience objects. Thus, "space" for Kant is not something existing – as it was for Newton. Space is an "a priori" form that structures our perception of objects in conformity to the principles of the Euclidean geometry. In this sense, then, this geometry is the correct geometrical structure of the world. It is necessarily correct because it is part of the "a priori" principles of organization of our experience.[7]

The Kantian claim about the a priori nature of Euclidean geometry is exactly what Poincaré mostly criticized.[8] He thought that our knowledge of the physical space is inferred directly from perceptions. This knowledge is a theoretical construct, that is, it is an explanatory hypothesis we infer from perceptions. The hypothesis provides an account for the regularity we experience. Its inferential nature is in direct opposition to the role of necessary, "a priori" principle it plays in the Kantian worldview. Although Poincaré does not endorse an empiricist account, he thinks that an empiricist view of geometry would be more adequate than the view held by Kantian scholars. Indeed, the idea that only a large number of observations can establish the correct geometry of the physical world sounds to him more plausible.

However, according to Poincaré, the empiricist approach fails to capture the true nature of geometry as well. As Sklar says, "nevertheless the empiricist account is wrong. For, given any collections of empirical observations, a multitude of geometries, all incompatible with one another, will be equally compatible with the experimental results" (Sklar, 1977, page 89). This quotation above illustrates the Poincaré idea that any hypothesis about the geometry of physical space is underdetermined by experimental evidence. Note well that the underdetermination is not due to our ability to collect experimental facts. No matter how rich and sophisticated are our experimental procedures for accumulating empirical results, the latter will be never enough compelling to pick the only correct hypothesis from the multitude.

Moreover, according to Poincaré, things seem actually worse than that: empirical results do not support at all any reason to think there might be a correct hypothesis in the multitude. Poincaré thought that this problem was grist to the mill of the conventionalist approach to geometry. The adoption of a geometry for physical space is a matter of conventional choice.

The Poincaré parable mentioned in Chapter 1 unravels this point quite unambiguously.[9] As I said there, Poincaré describes an imaginary two-dimensional world in which two mutually exclusive hypotheses of an Euclidean and hyperbolic geometry turn out to be equally consistent with the same collection of empirical data.

Now, in Poincaré's story the physical space of that imaginary world does have an actual geometry, which is the Euclidean one. But in virtue of some hidden dynamics affecting rods lengths in function of a specific spatial variation of the temperature, the inhabitants of the disk get out of measurements and of dynamical generalizations a manifest geometry which is an infinite Lobachevskian plan.[10] Indeed, the inhabitants are unaware of the peculiar temperature distortion of their rods, i.e., that at each point in space their rods all stretch (or all dilate) proportionally to the temperature's value at that point. The story also tells that at some point a group of inhabitants proposes a different interpretation of the collected data. The alternative hypothesis claims that the same empirical data also support a different model of the world's structure, that is, an Euclidean disk equipped with fields shrinking or dilating lengths.

Now, in the Poincaré parable, only one of the competing hypotheses is assumed to be the truthful one. This narrative feature should not suggest that the story conveys an argumentative line about the problem of epistemic access to the truth. Indeed, moving from our superior perspective to that of the disk's inhabitants would collocate us in the exact same situation. Indeed, it is a story about the impossibility to decide, even looking at comprehensive collections of empirical data, what of the two competing geometries is the correct one. The thing is that any arbitrarily large amount of empirical data, never mind how much comprehensive and exhaustive they are, cannot refute, or unambiguously supports, any hypothesis about the correct geometry of the world.[11] In order to bypass this impossibility *in principle*, Poincaré argues for a conventionalist solution to the problem of establishing the epistemological status of geometry.

Now, my reading of Poincaré extrapolates from his parable a lesson about fundamental quantum physics. In Chapter 1, the analogy between the parable and the physical scenario of underdetermination is presented in the case of a general classical Hamiltonian, and then extended to a general quantum Hamiltonian. Here, I am restricting my attention to the quantum strings dynamics and to the existence of T-duality that characterize them. The problem of underdetermination of geometry by physics involves geometrically inequivalent extra dimensions.

One may claim that inserting the quantum perspective into the Poincaré story adds some conceptual dimensions to the latter. On the one side, once the analogy is extended to the quantum string action, the separation between the manifest image of the world and its actual fundamental stringy structure becomes a divorce between a low-energy description of the world, containing geometry, and a high-energy one which does not contain geometry – indeed, geometry does not need to be there in virtue of its lack of explanatory power. In this new setting, the two descriptions become two incommensurable images of physically different levels of reality – different in virtue of the different physical length scales and parameters defining them. So, here the scenario of incommensurably different hypotheses replaces Poincaré's original one in which the two hypothesis are competitors.

On the other side, in the context of T-duality, the strings dynamics do not appear to be affected by geometrically inequivalent modifications of the extra dimensions. Although the geometrically inequivalent descriptions share the same physical length scale, still they are not competitors in virtue of their dynamical (or physical) equivalence. Differently, in the original Poincaré's story the two hypotheses facing each other are geometrically inequivalent, but also dynamically (or physically) inequivalent, despite they are confirmed by the same empirical data. Indeed, when the second group of inhabitants proposes the hypothesis of a Euclidean disk equipped with dynamical shrinking/dilating fields they are not only proposing a different geometrical model, but also a different dynamical one.

4.3 T-duality. Mathematical features and philosophical implications

Let's repeat the main feature of T-duality. Two string theories differing from each other in virtue of the geometrical inequivalence of their respective geometrical arenas may turn out to have the very same physical content. The geometrical differences considered here are a subset of a larger set of geometrical inequivalence involving the compact extra dimensions K of the stringy higher-dimensional arena $S \times K$. This subset of geometrical equivalence identifies the string physics' insensitivity described as T-duality.

Now, what does T-duality say about the nature of these higher-dimensional arenas in the theory? As I will argue, it says they are emergent. Here we consider a simplified case of T-duality by only considering the bosonic string. Since compactness of the extra dimensions plays a central role in T-duality, in Section 4.3.2, I will explain what compactness means in this context. Heuristically speaking, it is a topological property. In string theory, the "action" of compactifying one or more extended dimensions of some arena means wrapping each of them in a circle. The transformation induced by that "action" entails a change of the traditional formal language used to describe those dimensions. In this new theoretical setting, extra dimensions can only be formally represented by periodic variables – with period 2π times the circle's radius each gains.[12] Then, the fourth and fifth subsections present some cases of bosonic compactification and bosonic T-duality. For now, let's start with some preliminary remarks about duality and compactification procedures.

4.3.1 *Some preliminary remarks on dual theories: what does "same physical content" mean?*

The notion of duality should not be confused with that of physical symmetry and of gauge symmetry.[13] Physical symmetries in general (and in particular for the bosonic string) map distinct physical states to one another. An orbit produced by their group action over the Hilbert space $\mathcal{H}$ (which for the bosonic string is the space of all its possible physical states) is made of points, each of them representing different physical situations.

Gauge symmetries map physical states to one another as well, but both the transformed and untransformed states represent the very same physical situation. This fact explains why gauge symmetries are usually considered to be unphysical, that is, to be simply features of the way in which the theory represents the same physical situation. So, such symmetries are traditionally considered to be a vehicle of descriptive redundancy. The latter may be swept away taking the quotient of the Hilbert state $\mathcal{H}$ with respect to the action of the gauge group.

Now, if we redefine the notion of a physical state in terms of being a gauge orbit, then the difference between physical symmetries and gauge symmetries can be restated in the following: physical symmetries map gauge orbits to one another, whereas gauge symmetries are transformations inside gauge orbits.[14]

Duality is a type of symmetry that involves a space whose elements are not physical states. Both the transformed and untransformed objects are physical theories. Two distinct physical theories are dual of one another if they produce the same physics, that is, if they have the same physical content. Let's see more closely this notion of same physical content.

Now, what does it mean to say that the two theories have the same physical content?

Dean Rickles in his *A Philosopher Looks at String Dualities* made one of the first philosophical attempts of understanding this peculiar intertheoretical relation. His work became later on the standard philosophical way of presenting dualities.[15] Two theories have the same physical content when they make precisely the same predictions about all observable phenomena, that is, when the expected values of any observable in any state are the same in both theories. One may say that this characterization relies on some old verificationist idea according to which the physical content of a theory is entirely defined by the complete set of the theory's observables. So, any attempt to define what duality between theories means in this context would seem to require that their complete sets of observables must perfectly match with one another.

However, endorsing any form of verificationism is not really a mandatory requirement here to characterize dual theories. Indeed, the constraint dictated by duality on the observables of each theory of any dual pair does not require that any observable of each theory must correspond to some observable phenomenon. One might well say that two dual theories must have the same expectation value for each observable *that can be "observed"*. Any physical theory has also a "representational part," that is, a subset of observables which are not "observable." In this sense, the physical content of a theory is that part of the theory that feeds its ontology, and it might outrun the set of the theory's observables.

Now, dual theories can appear to be completely different, that is, all observable phenomena they equally predict might be described by different mathematical language, within different physical scenarios (i.e., scenarios depicting physical processes occurring at very different length scales). For example, a dual couple can be made by theories having different spacetime geometries or different number of dimensions. It can be made by two theories such that one is quantum but the other is not, one strongly coupled and its dual weakly coupled, and so on.

Now, a useful feature of dualities is that the idea of physical equivalence they deliver turns out to be a powerful tool for computational purposes. What does that mean? The simple setting presented by Brian Greene in his *The Elegant Universe* can clarify this point.[16] Let's assume we are trying to calculate the value of some physical property, like some force charge or some particle mass, within the mathematical/physical framework of some theory. Let's assume also that there are some technological obstacles to experimentally testing the computational result. Therefore, we can only rely only on other mathematical predictions about what physical result we may likely obtain. However, it might be the case that at some point, even these additional mathematical computations turn out to be out of control.

Now, the intractability of the problem might be bypassed if the theory in which we are trying to compute the value of a physical property has a dual partner in which things work out, that is, if there is a mathematically and ontologically different theory, still physically equivalent, in which computations of the same physical property may be performed by a simpler mathematical machinery. Of course, it may turn out that within the dual theoretical framework, computational tools are even more difficult.

However, the important point about the good side effects of having a dual pair of theories is that duality makes the correspondence between theories an exact one. It is in virtue of this exactness that one can translate results from one side to another without using approximations between theories.[17] In other words, duality has nothing to do with traditional approximation of one theory by another. Indeed, any theory's approximation would not preserve the exact physical content of the theory that gets approximated. Moreover, duality has nothing to do with the notion of redundancy delivered by gauge symmetries. The differences between two dual theories are not purely representational; rather they separate the main ontological cores of the two theories, along with their formal articulations.

Let's conclude this subsection with a slightly more rigorous presentation of dualities that prepares the ground for what follows in this chapter and in the remaining ones. The definition of duality I am presenting here is extrapolated from *Geometric Physics* by Vafa (1998, pages 539–540). Many contemporary theories in physics (especially quantum gravity theories) appear to have ubiquitous features at the interface between pure mathematics, physics, and metaphysics. Often crucial features of the theory's physical content are derived from the features of its formal articulation. Then, these formal features turn out to have a philosophical significance. An example of how formal features acquire philosophical meaning is the one presented in Chapter 3: reimposing conformal symmetry of the quantum string's action produces a formal derivation of the physical solutions of the Einstein field equations. That derivation becomes a philosophical notion once it is read in light of supervenience-based emergence.

Now, contemporary physical theories are sometimes individually associated to abstract mathematical objects encoding in simpler and more intuitive terms their formal and physical features. This sort of object, imported from advanced pure mathematics, is commonly called the moduli space of the theory. As I said, quantum string theory is often associated to a moduli space. Within the string theorists' circles, this abstract space is used for a variety of reasons. In some cases, its formal structure is built according to the epistemological need of classifying the several types of string theory formulated so far. In this case, the hope is that of finding the "fine moduli space" of the theory, that is, the moduli space that parameterizes and encodes some unique fundamental formulation of string theory, the one from which every other existing formulation can be gained by means of dualities. This logic mimics the one originally underlying the use of moduli spaces in algebraic and differential geometry, that is, to find a "fine moduli space" parameterizing some universal, general family of some type of geometrical objects, one from which any other less general family of the same type of geometrical objects can be obtained. In some other cases, the moduli space's formal structure has been suiting the need of studying the cosmological implications of the theory – like for example the multiverse scenario. The "space" I will construct in Chapter 6, only partially overlapping with these two formal apparatuses, is more similar to some kind of global "positioning" system. But different from any ordinary system like that, what it is "positioning" is the degree of background independence of string physics. For now let's stick to the presentation made by Vafa. Broadly speaking, contemporary quantum physical theories usually describe physical systems dynamics as depending on a number of physical parameters, whose different variation ranges produce different physical scenarios. One might say that a set of parameters, along with some of their variation intervals, identify a theory. Still loosely speaking, these physical parameters, if somehow glued together, might form the moduli "space" M associated to the theory.[18]

Now, let $\{O_{\alpha_i}\}$ be the complete set of observables of a quantum theory T describing the behavior of some quantum system Q. The set of observables of a quantum theory are simply the quantum physical properties of the systems studied by the theory. Loosely speaking, the values taken by these quantum properties are usually computed by correlation functions, that is, the canonical way of computing any observable's expectation value of any quantum system amount to be the computation of values taken by these correlation functions.

Less canonically, the correlation functions are also defined over the theory's moduli space in the following way: given the system Q, once the set of physical parameters, say λ, connected to the theory and to the system are chosen, for every $\lambda \in M$ and for every n, that is, for every set of observables, $\{O_{\alpha_1}, O_{\alpha_2}, \ldots, O_{\alpha_n}\}$ of the system Q, we have

$$< O_{\alpha_1} \ldots O_{\alpha_n} >= f_{\alpha_1 \ldots \alpha_n}(\lambda), \tag{4.1}$$

where the right side of the identity shows the observables' expectation values. Although this is just a simplified presentation, still it allows to get a first impression on how the parameters λ (that in Chapter 6 will gain a precise identikit) connect to the system's correlation functions. It is basically in virtue of this connection (and of something more as I'll show in Chapter 6) that these parameters bring the system's dynamical features into the topological structure of the moduli space.[19]

Now, the Vafa's definition of duality uses the language of moduli spaces. Let's denote a physical theory T describing the dynamics of a physical system Q with $T = Q[M, O_{\alpha_i}]$.

Definition 4.3.1. Two distinct physical theories $T = Q[M, O_\alpha]$ and $T' = Q'[M', O_\beta]$ are considered to be dual to one another if

1. the two parameters moduli spaces M and M' are isomorphic and
2. there is an isomorphism between O_α and O_β compatible with all the correlation functions.

In other words, T' is dual to T if there exists an isomorphism h

$$M \leftrightarrow^h M'$$

$$\lambda \leftrightarrow \lambda'$$

such that for all $\lambda \in M \ \exists! \ \lambda' \in M'$ and for all α, there exists a unique β such that

$$f_\alpha(\lambda) = f'_\beta(\lambda').$$

Some features of the isomorphism between the two moduli spaces reveal information about what type of duality connects the two corresponding theories.[20] Basically, if the isomorphism between the two moduli spaces is trivial, then the duality relation amounts to be a case of self-duality, that is, a case in which the two apparently distinct dual theories are indeed the same one. In other words, the trivial isomorphism is actually an automorphism of the moduli space associated to the theory. An automorphism maps the space into itself. So, in this case, it relates internal regions associated with one formulation of the theory to those internal regions associated with the other theory's formulation. Differently, if the isomorphism between two moduli spaces is not trivial, then the corresponding duality relation is not self-duality and in general it relates very different theories.

4.3.2 *Some preliminary remarks on compactification*

Why are extra dimensions compact in string theory? Loosely speaking, compactification is the mathematical technique used to make their volumes tiny enough to be undetectable at low-energy regimes. Although crucial to increase the theory's predictive power, the assumption about the existence of extra dimensions also requires some extra move to preserve other aspects of the theory's empirical adequacy and to contain the increasing degree of abstractness. Indeed, one might say that the stronger predictive power is achieved at the cost of plausibility and concreteness. So, making the extra dimensions undetectable, smaller than the smallest length scales we can probe, sounds like an effective strategy of preservation.

Now, still loosely speaking, the feature of smallness combined with that of circularity produce the mathematical prototype of a compact structure. Then one may assume that at any point of the extended ordinary dimensions, small curled up extra dimensions live without being seen. To say something more precise about compactification of dimensions in string theory, I introduce two equivalent mathematical definitions of compactness.[21] The first one is applicable to sets (or spaces) which are purely topological, whereas the second one is applicable to sets (or spaces) which are also geometrical sets (or metric spaces), that is, spaces also equipped with a geometrical structure, in addition to a topological one. In what sense the existence of a topological structure might not entail the existence of a geometrical one?

Broadly speaking, a geometry defined on a set usually defines a relational network among the parts of the set which is usually stronger than that defined by a topology. "Stronger" here means having the property of satisfying a bigger number of axioms and constraints. This fact implies that a geometrical structure defined on a set produces a richer type of relations among its parts. Indeed, when a space (or set) is provided with both topological and geometrical structures, it is almost always the case that the topology is induced by the geometry (or metric) in virtue of the fact that the topology obeys a smaller number of axioms and constraints than those obeyed by the geometry. For example, in a topological space without a geometry (or metric), you cannot define the distance between two points, that is, a precise measure of their degree of closeness. But you can define a weaker notion of closeness like that of neighborhood of any point. But the relevant thing is that not all topologies one may define on a set are derived from a geometry (or metric). And these topologies are going to be very important for our argument in favor of the idea that quantum string theory does not posit any fundamental geometry.

Now, let's introduce the two definitions:

Definition 4.3.2. A topological space (or topological set) S is compact if and only if every open cover of S has a finite subcover.

If a topological space is also a metric space, then the definition above can be proved to be equivalent to the following definition:

Definition 4.3.3. A topological space (or topological set) S is compact if and only if it is closed and bounded,

where one may say that a metrical set is "closed" just in case for any metrical sequence made out of the set's elements, the sequence converges to a finite limit and the latter is still an element of the set. Moreover "bounded" denotes the property of being a metrical set of finite size, although having infinite, uncountable cardinality.

Both definitions do not contain any explicit reference to the property of "being circular." However, in the string physics circles, "being compact" is often used with the additional meaning of "being curled up or being circular." This extra meaning turns out to be quite close to the original mathematical one though. Indeed thinking of compactness in terms of being curled up is an intuitive way of representing some sort of finiteness of the space's size. Moreover, the mathematical procedure of getting compactness requires the "action" of curling up something flat by curving up the edges. For example, one can think of the circle in terms of an open line equipped with a specific function of identification between pairs of points.[22] In other words, if one identifies points on the open line separated by a distance equal to $2nR$, then one gets a circle; that is, two points P and Q on an open line are considered to be the same point if their coordinates differ by an integer multiple of $2\pi R$:

$$P \sim Q \Leftrightarrow x(P) = x(Q) + 2n\pi R.$$

Strictly speaking, having a small volume is not necessarily entailed by the property of being compact. Indeed, compactness can be also a property of large and extended dimensions. For example, the three observable spatial dimensions of our universe have a visible extension of about 15 billion light-years. No astronomical observation can currently tell us what happens beyond that distance. They could either continue to extend indefinitely or curl up in the shape of a huge circle that cannot be seen with our current telescopes; that is, our familiar extended dimensions might be compact as well.[23]

As I said, positing compact extra dimensions produces a type of higher-dimensional background mathematically representable by a product space of the form $S \times K$ – where S is the four-dimensional general relativistic spacetime and K is the n-dimensional compact part. The latter is mathematically represented by a complex manifold. Here the choice of a word like "manifold" is not dictated by the endorsement of any form of substantivalism about spacetime. Indeed, in this context, the word "manifold" actually means topological structure, that is, it actually denotes a purely relational construct.

In this chapter, I argue in support of the idea that any type of compact extra dimension postulated by the theory has an emergent nature in the theory's physical scenario. In other words, this chapter contains a thesis of background independence of string theory's physical content with respect to the extra dimensions K. As I said, historically introduced for consistency reasons, that is, almost like instrumental tools, the compact extra dimensions soon acquired some robust metaphysical status in the theory, in virtue of the fact that they strengthen its predictive power. But again, nothing in the physical content of the theory supports their fundamentality. Rather, looking for some more fundamental structure in the theory's physical ontology, one encounters something different from the extra dimensions K, although somehow connected to them. More precisely, what indeed plays a crucial role in the research of fundamental physical features in the theory is the space of deformations of K, that is, the theory's moduli space mentioned above. As we will see in Chapter 6, the local structure of this abstract space is built by gluing together sometimes geometrically inequivalent deformations of K, all sharing the same topological invariants, some other times topologically inequivalent ones.

In this chapter, T-duality is the main core of the emergence thesis. The setting is a toy scenario, precisely a bosonic string theory over a circle of radius R, and hence with only one compact extra dimension. As we will see, bosonic T-duality is a physical equivalence between a string theory over a circle of radius R and another string theory over a circle of radius $1/R$. Once R is fixed, bosonic string physics is insensitive to a specific type of geometrical deformation of the extra dimension's radius.

4.3.3 *The history of compact extra dimensions: Kaluza-Klein method*

The strategy of introducing compact extra dimensions to increase physical theories' predictive powers did not originate in the context of string theory. However, the original motivation was pretty much the same as that used in the string theory circles, that is, the desire to find (via the posit of compact extra dimensions) some unifying framework from which apparently different physical theories might be derived, for some physical limit reaching some limit value.

Now, in the earlier 1920s, a Polish mathematician named Kaluza conjectured that, despite our ordinary perception of a three-dimensional spatial arena, the number of spatial dimensions in our universe might be bigger than three. As far as I know, Kaluza and the Swedish mathematician Klein, who refined Kaluza's initial proposal, have been the first scientists in the twentieth century to introduce the idea that the spatial fabric of our universe may have both extended and curled up dimensions. The rationale behind this conjecture was that of identifying possible intertheoretical relations between Einstein's general relativity and Maxwell's electromagnetic theory. Indeed, Kaluza at some point discovered that by adding one compact spatial dimension to the four-dimensional universe, the unification between the two theories was possible, hence providing a compelling field theory framework encompassing both. The mathematical language used was extrapolated from the formal articulation of field theory, which later on remained virtually unchanged in string theory – although in the string case few features of the original procedure get modified by some strings' peculiarities.[24] Here, I won't be analyzing in detail the Kaluza-Klein reduction. Nevertheless, a summary of few basics might enlighten the string version of their reduction.[25]

Kaluza-Klein reduction produces the unification of electromagnetism and general relativity by proving that the symmetry group of electromagnetism can be obtained from the five-dimensional compactified gravity. Loosely speaking, the latter is given by a five-dimensional spacetime equipped with a metric tensor G_{MN}, $M,N = 0,\ldots,4$ – the four-dimensional part is any of the general relativistic solutions, and the fifth dimension is the compact one. Using here the notations introduced in Section 4.1, this type of compactified spacetimes might have local structures representable as the product space $\mathbb{R}^4 \times S^1$. Using the natural coordinatization, the four coordinates in $\mathbb{R}^4$ are the ordinary Cartesian ones denoted by

x^μ, $\mu = 0, \ldots, 3$, whereas the fifth coordinate in S^1 is periodic, that is, $x^4 = x^4 + 2\pi R$, where R is the radius of the circle S^1.

Now, the five-dimensional coordinate transformation $x^\mu \longrightarrow x'^\mu = x^\mu + \epsilon^\mu(x)$ (where $\epsilon^\mu(x)$ is the partial derivative ∂x^μ) is an invariance of the five-dimensional theory. This claim won't be proved here – for detail about the proof, I refer the reader to Chapter 3 of Johnson (2003). This transformation is a conformal one if the metric is invariant up to some overall scale factor. Under this coordinate change, the metric tensor also transforms in the following way:

$$G_{\mu\nu} \longrightarrow G'_{\mu\nu} = G_{\mu\nu} - (\partial_\mu \epsilon_\nu + \partial_\nu \epsilon_\mu), \tag{4.2}$$

where $\partial_\mu \epsilon_\nu + \partial_\nu \epsilon_\mu = \partial G_{\mu\nu}$. Then, in order for this transformation to be conformal, it must be that $\partial_\mu \epsilon_\nu + \partial_\nu \epsilon_\mu = \Omega(x) G_{\mu\nu}$, where $\Omega(x)$ is a scale factor.

The metric $G_{\mu\nu}$ factorizes naturally into $G^5_{\mu\nu}$, G^5_{44}, and $G^5_{\mu 4}$, where the superscript reminds that the quantities considered are in five dimensions. From the four-dimensional perspective, $G^5_{\mu 4}$ is a vector. The five-dimensional coordinate transformation above is expressed in its most general form. For our purposes, I pick the specific instantiation characterized by $\epsilon_4(x^\mu) \neq 0$, and $\epsilon_\mu = 0$ for $\mu = 0, 1, 2, 3$. Under this specific change of coordinates, the vector $G^5_{\mu 4}$ transforms in the following way:

$$G^5_{\mu 4} \longrightarrow G'^5_{\mu 4} = G^5_{\mu 4} - \partial_\mu \epsilon_4(x). \tag{4.3}$$

Now, this transformation of the metric turns out to be a gauge transformation of the symmetry group $U(1)$ of electromagnetism. The result is given here without proof.[26] Loosely speaking, a gauge transformation of the symmetry group $U(1)$ in general acts in the following way:

$$A_\mu \longrightarrow A_\mu - \partial_\mu \Lambda(x),$$

where A_μ is any vector proportional to $G^5_{\mu 4}$. In other words, the vector $G^5_{\mu 4}$ would be the specific term replacing the general vectorial term A_μ when dealing with our specific case of five-dimensional metric transformation. Then, Maxwell electromagnetic theory is proved by Kaluza and Klein to be the result of the compactification of gravity, because the gauge field is a component internal to the metric, i.e., it is the vectorial component $G^5_{\mu 4}$.

Now, to sketch why the fifth component of any physical states is not visible from our four-dimensional perspective, one might choose a physical property of a five-dimensional field system and then examine what kind of functional dependence its fifth component has from the compact dimension radius or from the variation of energy scales. The canonical choice is that of the linear momentum p of a field system. Its fifth component is a scalar function of the periodic variable. Because of this periodicity (i.e., $x^4 = x^4 + 2n\pi R$), that component turns out to be a quantized quantity looking like the following:

$$p^4 = \frac{n}{R}. \tag{4.4}$$

Now, as I said, the theoretical framework within which the reduction is performed is that of a field theory in dimension five. By using the five-dimensional law of motion obeyed by these fields, one can find the infinite family of patterns of radiation. Loosely speaking, if fields have a non-zero fifth component p_4 of their momentum p, that component will produce an energy density along the periodic dimension equal to the following amount (Johnson, 2003, ch. 4):

$$-p^4 p_4 = \frac{n^2}{R^2}, \tag{4.5}$$

Then, one gets a tower of states that become heavier as R becomes smaller. In other words, the smaller R is, the heavier these states become. Their heaviness is responsible for making them invisible because it would take too much energy for light to travel from them to us.

So, at the end of the day, the smallness of the extra dimension's radius really accommodates phenomena, since it explains why the Kaluza-Klein are not visible from the four-dimensional perspective. Only physical states with properties vanishing along the fifth dimension are visible. Positing a big radius instead would produce an extended extra dimension, a one-dimensional dimension comparable to large spatial scales. In this case, the Kaluza-Klein states would become visible exactly for the same inversely proportional relation between momentum and radius used above, i.e., (4.5). So between the two, the posit of small radius is the most empirically adequate.

4.3.4 *Compactification in the closed string case*

This section deals with an extension of the Kaluza-Klein compactification process to the bosonic closed string. As I said in Chapter 2, the model requires 26 dimensions, 22 of which are compact – in what follows, D denotes the spacetime's dimension and sticks to the usual convention $D = 0, \ldots, 25$. In this chapter, the compactification process of the bosonic string is sketched within a simplified setting since only one of the extra dimensions gets compactified, i.e. the X^{25}.

Now, in Chapter 2, the formula (2.9) shows the easiest way to write string equations down. It is indeed a simple wave equation depending on the two string parameters σ and τ, whose solutions can be naturally written as superpositions of left-moving waves and right-moving ones –

formula (2.10). The reason for which I am referring here to this part of Chapter 2 is that the string total momentum functionally depends on string solutions to the string equations of motion, and like in the field case of compactification, the system's total momentum is a main tool of exploration of the consequences produced by compactness.[27]:

$$p^\mu = T\int_0^{2\pi} d\sigma \frac{dX^\mu}{d\tau}(\sigma) = \frac{1}{\sqrt{2\alpha'}}(\alpha_0^\mu + \tilde{\alpha}_0^\mu). \tag{4.6}$$

The right side of this identity contains the vibrational zero-modes of the string (lowest-energy vibrations). From the closeness of the string, it is possible to derive a precise relation between these two zero-modes. To this aim, one needs to move along the closed string, that is, to transform the string parameter σ via the following transformation:

$$\sigma \longrightarrow \sigma + 2\pi. \tag{4.7}$$

Since the oscillation terms are periodic in σ, this transformation induces the following one acting on the string solutions:

$$X^\mu(\sigma,\tau) \longrightarrow X^\mu(\sigma,\tau) + 2\pi\sqrt{\frac{\alpha'}{2}}(\alpha_0^\mu - \tilde{\alpha}_0^\mu). \tag{4.8}$$

Since any X^μ is still a one-value function (i.e., it's not periodic yet), we need to impose that

$$\alpha_0^\mu - \tilde{\alpha}_0^\mu = 0. \tag{4.9}$$

Then it follows:

$$\Rightarrow \alpha_0^\mu = \tilde{\alpha}_0^\mu. \tag{4.10}$$

Now, replacing this equality in the momentum formula (4.9), one gets

$$p^\mu = \sqrt{\frac{2}{\alpha'}}\alpha_0^\mu = \sqrt{\frac{2}{\alpha'}}\tilde{\alpha}_0^\mu. \tag{4.11}$$

The fact that compactness is not part of the story yet (we are just moving along the closed string) is indicated by the continuum spectrum of values taken by the momentum p^μ along any direction, hence in particular along X^{25} – the dimension chosen for compactification. In other words, this continuity means that the direction X^{25} is not periodic yet.

After compactifying X^{25}, that is, after performing the identification $X^{25} \cong X^{25} + 2\pi R$, the momentum along that direction turns out to be quantized, that is, the function p^{25} only takes discrete values:

$$p^{25} = \frac{n}{R}. \tag{4.12}$$

Although this formal scenario almost completely overlaps with that of field theory, the two do not match perfectly. Indeed, if one moves along the closed string, since X^{25} is now periodic (i.e., not a one-value function anymore), one gets a physical quantity not present in the field case, that is,

$$X^{25}(\tau,\sigma+2\pi)=X^{25}(\tau,\sigma)+2\pi\omega R. \tag{4.13}$$

The new quantity is ω. It is an integer, and it is called the closed string's *winding number*. In other words, a closed string can wind around the compact dimension. This type of behavior is not something that can be exhibited by fields or point particles.

So, in the string scenario, we have two equations:

$$p^{\mu}=\frac{n}{R}, \tag{4.14}$$

$$\alpha^{25}-\tilde{\alpha}^{25}=\sqrt{\frac{2}{\alpha'}}\omega R, \tag{4.15}$$

from which one gets an expression for the zero-modes along X^{25}:

$$\alpha_0^{25}=\left(\frac{n}{R}+\frac{\omega R}{\alpha'}\right)\sqrt{\frac{\alpha'}{2}}=\sqrt{\frac{\alpha'}{2}}p_L^{25}, \tag{4.16}$$

$$\tilde{\alpha}_0^{25}=\left(\frac{n}{R}-\frac{\omega R}{\alpha'}\right)\sqrt{\frac{\alpha'}{2}}=\sqrt{\frac{\alpha'}{2}}p_R^{25}.$$

One can use all these relations to compute the mass spectrum:

$$\begin{aligned}M^2&=-p^{\mu}p_{\mu}=\left(p_L^{25}\right)^2+\frac{4}{\alpha'}(N-1)=\\&=(p_L^{25})^2+\frac{4}{\alpha'}\left(\tilde{N}-1\right)=\\&=\frac{n^2}{R^2}+\frac{\omega^2R^2}{\alpha'^2}+\frac{2}{\alpha'}\left(N+\tilde{N}-2\right),\end{aligned} \tag{4.17}$$

where the moment contribution along X^{25} shows up separated from the total moment contribution along all the remaining dimensions. Also, N and $\tilde{N}$ are excitations levels of the left- and right- moving oscillators of the closed string.

Note that in the case of closed strings, the mass spectrum contains not only the tower of momentum states ($n\in\mathbb{N}$), like in field theory but also the tower of winding states ($\omega\in\mathbb{Z}$). The first term on the last line of (4.20) is the contribution given by the Kaluza-Klein's tower of momentum states in the string case (the compact moment), the second term represents the potential energy of the string which is winding, i.e., the term deriving from the tower of the winding states, which does not show up in the original Kaluza-Klein tower of states. Finally, the third one represents the usual excitation levels of the closed string oscillators.

Now, if $R \longrightarrow \infty$, then the mass formula shows that the winding states ($\omega \neq 0$) tend to disappear, being infinitely massive. Large radius entails large winding energy, whereas the momentum states become lighter, hence preferable from the energetic point of view. Therefore, the states characterized by $\omega = 0$ and $n \neq 0$ produce a continuum spectrum, describing in this way a physical configuration without compact dimensions. If instead $R \longrightarrow 0$, then momentum states tend to disappear. In ordinary field theory, the consequences of this fact would be simply that the remaining fields are independent of the compact coordinate. But in the string context, things are more complicated. Small values of the radius entail small winding energy. So, the pure winding states survive ($\omega \neq 0$, $n = 0$), and now they will be responsible for the continuum spectrum.

Now, some key facts about bosonic T-duality can be introduced. Let's consider a string configuration. First, physical properties are sensitive to the total energy of this configuration but not to the way in which the total energy splits into a vibration part and a winding part. Second, let's consider a large circular radius background for string propagation.

Based on what I presented above about the behavior of the spectrum triggered by the variation of R, I can claim that there exists a corresponding small circular radius background such that the string vibration energies are equal to the winding energies in the large radius background and such that the string winding energies are equal to the vibration energies in the same large one.

But, since the total energy is the same in both cases, there is no physical distinction between these two cases. Two bosonic theories referring to these two different geometrical backgrounds are empirically indistinguishable. More precisely, they are the same theory. In fact, each background of the pair can be thought as a result of slightly deforming the radius of the other one.[28] Shifting from one to the other means mapping a point over the theory's moduli space onto another point of the same moduli space. Bosonic T-duality is a toy duality, so it is not surprising to find out that it actually amounts to be a case of self-duality.

4.3.5 T-duality for closed strings

As we saw in the previous section, the closed string spectrum is invariant under the following transformation[29]:

$$T : n \leftrightarrow \omega, R \leftrightarrow R' \equiv \frac{\alpha'}{R}. \tag{4.18}$$

This symmetry is called T-duality. The compactified string theory over a circle of radius R has the same physical content as the compactified string theory over a circle of radius R', which is obtained by switching winding and moment numbers.

Given T-duality, we obtain the following transformations for zero-modes:

$$\alpha_0^{25} = \left(\frac{n}{R} + \frac{\omega R}{\alpha'}\right)\sqrt{\frac{\alpha'}{2}} \to \left(\frac{\omega}{R'} + \frac{nR'}{\alpha'}\right)\sqrt{\frac{\alpha'}{2}} = \left(\frac{\omega R}{\alpha'} + \frac{n}{R}\right)\sqrt{\frac{\alpha'}{2}}, \quad (4.19)$$

$$\tilde{\alpha}_0^{25} = \left(\frac{n}{R} - \frac{\omega R}{\alpha'}\right)\sqrt{\frac{\alpha'}{2}} \to \left(\frac{\omega}{R'} - \frac{nR'}{\alpha'}\right)\sqrt{\frac{\alpha'}{2}} = \left(\frac{\omega R}{\alpha'} - \frac{n}{R}\right)\sqrt{\frac{\alpha'}{2}},$$

that is,

$$p_R^{25} \longrightarrow -p_R^{25}, p_L^{25} \longrightarrow p_L^{25}. \quad (4.20)$$

In this simple case, when we apply T-duality to the theory of radius R (to get the theory of radius R'), the transformation switches the left and the right parts of the field $X^{25}(f(\sigma,\tau), h(\sigma,\tau))$, the one decomposable in $X_R^{25}(f(\sigma,\tau)) + X_L^{25}(h(\sigma,\tau))$. The switch is the following:

$$\begin{aligned} X^{25}(.,.) &= X_R^{25}(f(\sigma,\tau)) + X_L^{25}(h(\sigma,\tau)) \Rightarrow^{T\text{-}dual} X'^{25}(.,.) \\ &= X_L'^{25}(f(\sigma,\tau)) - X_R'^{25}(h(\sigma,\tau)), \end{aligned} \quad (4.21)$$

where $f(\sigma,\tau)$ and $h(\sigma,\tau)$ are functions of the string parameters taking complex values. The T-duality transformation also maps the components along X^{25} of all the remaining modes in this way:

$$\alpha_\mu^{25} \longrightarrow \alpha_\mu^{25}, \quad (4.22)$$

$$\tilde{\alpha}_\mu^{25} \longrightarrow -\tilde{\alpha}_\mu^{25}.$$

Therefore, the field X'^{25} keeps sharing with X^{25} the same energy-momentum tensor. The physical quantities expressed by the correlation functions of both theories will be the same. Hence, the theory of radius R and the theory of radius $\frac{\alpha'}{R}$ are physically equivalent.

Now, one can see how the extended nature of strings, via the parameter α', plays a crucial role in determining T-duality. The parameter is connected to the string tension T:

$$\alpha' = \frac{1}{2\pi T}. \quad (4.23)$$

Also, α' fixes the minimum distance. The minimum distance scale one might "see" according to the theory is $x \sim \sqrt{\alpha'}$. When $R = \sqrt{\alpha'}$ in (4.21), $R' = R$. So, $R = \sqrt{\alpha'}$ seems to be the minimum radius: if you shrink R below that value, you will get a theory for large radii.

4.3.6 *T-duality for open strings*

The setting is the same as that of the closed string: the only direction compactified is X^{25}, that is, $X^{25} \cong X^{25} + 2\pi R$. The compactification of X^{25}

again transforms the momentum along that direction in a quantized momentum, $p^{25} = \frac{n}{R}$, with $n \in \mathbb{Z}$. However, open strings cannot wrap around the periodic dimension. Therefore they don't have a winding number around that dimension. Since $\omega = 0$, the mass spectrum formula appears to be slightly different.[30]:

$$M^2 = \frac{n^2}{R^2} + \frac{4}{\alpha'}(N-1). \tag{4.24}$$

Along a similar line of thought, when $R \longrightarrow 0$, those states with non-zero internal momentum get an infinite mass. For this reason, they disappear. However, different from the case of closed strings, any continuum spectrum of states deriving from the winding tower won't be found. Therefore, the remaining fields will be independent of the compact dimension and we'll have a theory with one dimension less than the initial theory. Indeed, in this limit, the compact dimension is lost.

This fact might sound contradictory. Indeed (interactive) open string theories must contain closed strings. But, as I said, the limit $R \longrightarrow 0$ in the case of closed strings produces a D-dimensional theory, whereas the same limit in the case of open strings produces a $(D-1)$-dimensional theory. In spite of the non-fundamental nature of dimensionality in string theory, this dimensional mismatch seems to produce an inconsistency. Note well though that the limit $R \longrightarrow 0$ in the open case produces a constraint that only applies to the endpoints of the open string. That is because the internal part of an open string cannot be distinguished from that of a closed string. Whether the string is open or closed, the internal part vibrates according to the same patterns. So, there must be something about the dynamics of the endpoints that produces a difference. To understand what this difference amounts to be, one needs to see how T-duality works in this case and what kind of information is contained in the boundary conditions concerning the endpoints' vibrational modes.

Now, the expansion mode of an open string along the compact dimension X^{25} decomposes in the following way:

$$X^{25}(f(\sigma,\tau), h(\sigma,\tau)) = X^{25}(f(\sigma,\tau)) + X^{25}(h(\sigma,\tau)). \tag{4.25}$$

Let's assume the string's endpoints are free to move. T-duality in the open string case works generally in the following way:

$$\begin{aligned} X^{25}(f(\sigma,\tau)) &\longrightarrow X^{25}(f(\sigma,\tau)), \\ X^{25}(h(\sigma,\tau)) &\longrightarrow -X^{25}(h(\sigma,\tau)). \end{aligned} \tag{4.26}$$

Then, applying T-duality to the components of the expansion mode above, one gets

$$\begin{aligned} X^{25}(.,.) = X^{25}(f(\sigma,\tau)) + X^{25}(h(\sigma,\tau)) &\longrightarrow X'^{25}(.,.) \\ = X^{25}(f(\sigma,\tau)) - X^{25}(h(\sigma,\tau)). \end{aligned} \tag{4.27}$$

Therefore, by writing explicitly the T-dual expansion mode, one gets

$$X'^{25}(.,.) = X^{25}(f(\sigma,\tau)) - X^{25}(h(\sigma,\tau)) \tag{4.28}$$

$$= x'^{25} + 2\alpha'\frac{n}{R}\sigma + i\sqrt{2\alpha'}\sum_{n\neq 0}\frac{1}{n}\alpha_n^{25}e^{-in\tau}sin(n\sigma).$$

In the sector of zero-vibrational modes along X^{25} – that is $x'^{25} + 2\alpha'\frac{n}{R}\sigma$ – there is no dependence on the worldsheet's coordinate τ. Since the momentum of the string mass center is a function of τ, it turns out to vanish.

Interestingly, this fact means that the extremities of the T-dual open string, namely, $\sigma = 0$ and $\sigma = \pi$, don't move along the direction X'^{25}. Therefore, the T-dual string turns out to be fixed, i.e.,

$$p'^{25} = T\int_0^{\pi} d\sigma\partial_\tau X'^{25} = 0. \tag{4.29}$$

So, T-duality basically changes boundary conditions, that is, it changes the conditions relative to the expansion modes of the two extremities of the T-dual open string. In this case, the Neumann boundary conditions on the extremities of the untransformed open string, that is, $\partial_n X \equiv \partial_\sigma X = 0$, have been replaced by the Dirichlet boundary conditions on those of the transformed one, that is, $\partial_t X \equiv i\partial_\tau X = 0$, where n is the direction normal to the boundary and t the tangent one.[31] In addition to this switch, based on the derivative showing up in the Neumann boundary condition, one understands that if the extremities of an open string can move, they are confined to live on a hyperplane. In this case, the extremities lie on the same hyperplane in the periodic T-dual space:

$$X'^{25}(\tau,\pi) = X'^{25}(\tau,0) + n2\pi R'. \tag{4.30}$$

However, other configurations in which the extremities move (or are fixed) on different hyperplanes are possible. In any case, what is important to keep in mind is that the extremities of any open string move with $D-1$ degrees of freedom.

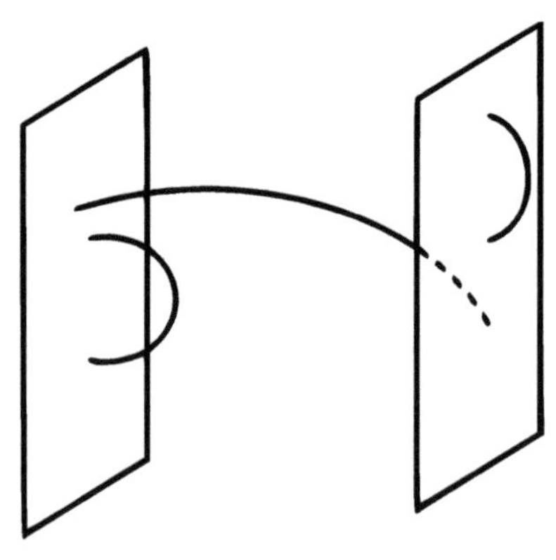

Now, the type of hyperplanes onto which open strings' ends attach turns out to play a crucial role in string theory as of the second superstring revolution that took place in the early 1990*s*. They are called *D-branes*. The latter will be introduced and described in the next chapter because of their roles in the holographic duality. Although they populate string theory's physical ontology, I will argue that they are not fundamental in the theory. Indeed, one might argue that D-branes are derived structures; that is, they are not put in the theory by hand. Quantum mechanically they appear to be made out of composite string excitations becoming visible only at strong coupling regime.

Although being derived structures, D-branes are very important in string theory. It is believed that quantum fields described by Yang-Mills theories (for example electromagnetism) involve strings that are attached to D-branes. This idea might explain why the quanta of gravity (gravitons) are not attached to D-branes. They can travel through a D-brane, and that would explain why we cannot see them. Our universe might be a boundary of some higher-dimensional bulk. The interactions in our world are mediated by particles that according to string physics might be seen fundamentally as strings stuck to the boundary brane. Gravity is mediated by strings that can leave the brane and travel away from it into the bulk. That would also explain why gravity is a much weaker force than the electromagnetic one.

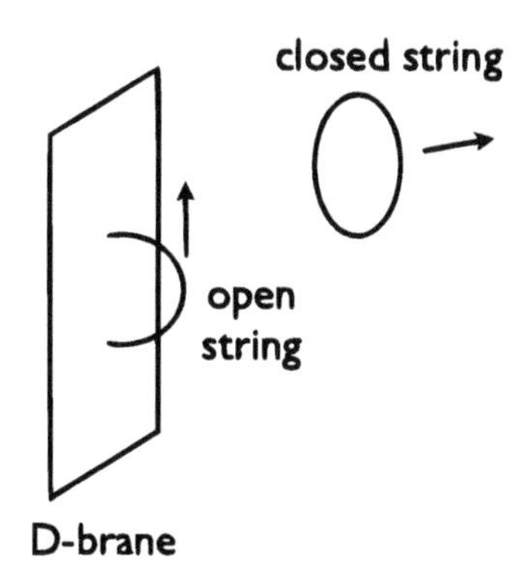

Notes

1 Although I am trying to present what I think is the original formulation of the Poincaré problem, I won't be concerned here with sophisticated exegetic issues. My reconstruction of the Poincaré view on the status of geometry heavily relies on that proposed by Sklar in his "Space, time and spacetime" – see Sklar (1977, section F, pages 89–103). Sklar attributes to Poincaré a conventionalist position. Whether or not this is the most faithful interpretation of Poincaré view is left aside here. What makes me choose Sklar's reading among other historical interpretations of the 19th century French philosopher, is that it definitely helps to illuminate the important philosophical points.

2 For a general discussion on Duhem's presentation of the problem, see http://plato.stanford.edu/entries/scientific-underdetermination/, whereas for a more deep analysis of his work, see his "The Aim and Structure of Physical Theory," trans. P. Wiener, Princenton Science Library, 1991. Here I won't be concerned with analyzing Duhem's position.
3 For a general discussion on Quine's position, see http://plato.stanford.edu/entries/scientific-underdetermination/, whereas for a detailed analysis of his position, see his "On Empirically Equivalent Systems of the World," *Erkenntnis, 9 : 313--328, 1975. Here I won't analyze Quine's view on this issue.*
4 See Sklar (1977, section *F*, pages 89—90).
5 The answer to the question is grounded on my interpretation of the following passages of the First Critique: Kant (1998, A 39/B 56).
6 See Kant (1998, A 23/B 38).
7 See Kant (1998, A 39/B 56).
8 See Sklar (1977, section *F*, pages 90).
9 See Sklar (1977, pages 92, 93).
10 The two-dimensional disk is heated to a constant temperature at the center, whereas along the radius *R*, it is heated in a way that produces a temperature's variation described by $R^2 - r^2$. Therefore, the edge of the disk is uniformly cooled to 0^0.
11 See Sklar (1977, pages 93–97).
12 The mathematical procedure of compactification is presented in Section 4.3.2.
13 See Rickles (2010, pages 54–56).
14 See Rickles (2010, page 55).
15 see Rickles (2010, pages 63–64).
16 See Greene (1999).
17 Rickles (2010, page 56).
18 How we should think about these gluing conditions in the case of string theory will be one of the main issues unraveled in Chapter 6. Let's just say here that this "space" encodes multilayers information about string theory's physical content in virtue of the fact that its topological structure is produced by those gluing conditions. The moduli space's topology, along with a specific fiber bundle I will define on top of it, carry information about the degree of background independence of the theory's physical content.
19 More on this in Chapter 6.
20 See Vafa (1998), and also Rickles (2010).
21 Also, see http://en.wikipedia.org/wiki/Compact_space for more detail, still introductory, about these two definitions.
22 See Zwiebach (2009, pages 31–32).
23 See Greene (1999, page 248).
24 See Greene (1999, pages 186–192).
25 This presentation relies on *D-branes*, by Clifford Johnson, see Johnson (2003, ch. 4).
26 For detail on the proof, see Johnson (2003, ch. 4).
27 In this section and in the next one, I unpack the presentation given by *BBS* in Becker, Becker, and Swartz (2007, section 6.1, pages 188 – 192.
28 More on this in Chapter 3.
29 Again, see Becker, Becker, and Swartz (2007, section 6.1, pages 188–192(for bibliographic references on this part.
30 In this section, I unpack section 6.1, pages 192–195 in Becker, Becker, and Swartz (2007).
31 Summarizing, the boundary Neumann conditions on the original open string's endpoints become the boundary Dirichlet conditions on the dual open string's endpoints and vice versa.

5 Holographic duality and emergence

5.1 Introduction

In the previous chapter, I used T-duality to argue that string theory admits emergent compact extra dimensions. In this chapter, I argue for a similar claim of emergence, but using a different physical duality, namely, the holographic one – also called the anti-de Sitter (AdS)/conformal field theory (CFT) correspondence. Part of the chapter's content will rely on a previous paper of mine on the metaphysics of dualities (Vistarini, 2016) where the AdS/CFT duality is analyzed along with a second case of holographic duality called the de Sitter (dS)/CFT correspondence, which is actually still a conjecture of duality.[1]

Broadly speaking, the AdS/CFT duality shows the redundancy of any posit about fundamental geometry in string theory more radically than T-duality. Indeed, even if one posits the geometry of one or the other side of the correspondence as fundamental, that geometry won't account for the physical content shared by the two theories (on each side, geometry is defined by some general relativity solutions combined with the extra dimensions). An attempt of philosophically understanding spacetime emergence via the holographic duality requires to unpack the physics and the formal articulation of the correspondences. But it also requires some more digging into the philosophical debate about emergence.

The AdS/CFT is a duality between a string theory over an AdS spacetime (the bulk spacetime containing gravity) and a CFT over its boundary (not containing gravity). This proved correspondence delivers a scenario in which the appearance of some spatial dimensions in the bulk is closely related to physical dynamics occurring in the boundary (not containing those spatial dimensions). The AdS/CFT conjecture was introduced in the early 1990s and proved few years after. What does "proof" mean in this context? Loosely speaking, proving a correspondence like the AdS/CFT duality means showing that if string theory admits as a physically possible solution some compactification of the AdS spacetime (i.e., a higher-dimensional spacetime obtained by adding to the AdS structure some compact extra dimensions), then its physical content against this higher-dimensional spacetime is indistinguishable from that of a quantum field theory living on the boundary of such spacetime. Once this is proved, the second crucial part of the proof consists in proving that string theory actually admits some compactification of the AdS spacetime.[2] As a matter of fact, an initial conjectured physical equivalence between an AdS theory with gravity and a CFT without gravity – dated back to the mid-1990s – has been successfully tested via a concrete instantiation in which a quantum string theory with gravity sits on the AdS side of the duality.

The two sides of the duality show theories positing not only geometrically and topologically inequivalent spacetimes but also incommensurably different physical ontologies. On the CFT side, there are no strings at all, whereas the AdS gravity theory is a quantum string theory. Note well that the inter-theoretical correspondence is an exact one in virtue of the fact that it is a duality. So, one might be facing here a twisted scenario of emergence, that is, the physical ontology of the quantum string theory on one side and that of the quantum field theory on the other might well be both not fundamental. This point will be unpacked in the final part of the chapter.

Finally, the holographic duality includes in the physical scenario a relatively new type of objects, namely, D-branes. They became part of the string theory's physical ontology dated back to the 1990s during the second super string theory revolution. D-branes have been just mentioned in the previous chapter as I was analyzing T-duality. D-branes there accounted for the peculiar behavior exhibited by the extremities of open strings, that is, the fact that these endpoints move with less degrees of freedom than the points situated in the internal part of open strings. The explanation is that they are stuck on these hyperplanes. However, D-branes appear to become central in the context of dualities when considering the non-perturbative physical scenarios delivered by the AdS/CFT duality.

What exactly are D-branes in string theory? Are they fundamental in the same way in which strings are? Are they also dynamical structures? Their ontological status in string theory is actually a matter of open debate. Two opposite views face each other. One assigning to them a fundamental status, similar to that enjoyed by strings, the other assigning a derived nature. In Vistarini (2016), I argue in favor of the thesis that D-branes are derived dynamical structures showing that they are not put in the theory by hand.

However, the two views are not necessarily in conflict. The lack of tension appears clearly if one makes a preliminary differentiation. What I mean is that from a perturbative point of view, that is, considering weak string coupling, as soon as we look at the D-branes' quantum mechanical structure, strings are unambiguously more fundamental than them. Indeed, within this physical scenario, D-branes appear to be specific configurations of classical fields somehow emerging from quantum strings. A useful analogy might be with magnetic monopoles, which in field theory are seen as classical configurations arising from the "more fundamental fields." Otherwise, when the string coupling is large enough that we can no longer trust perturbation theory, D-branes appear to be "equally fundamental" as strings. But that does not entail that they belong to the fundamental physical ontology of the theory. What that means is that when the string coupling becomes strong, some D-branes may become much lighter than they would be in the weak coupling regime; hence, they become potentially detectable in the same way in which strings are. Then, by including strongly coupled regimes, one finds what Zwiebach and Taylor call a "brane democracy" (see Taylor and Zwiebach, 2004).

D-branes are dynamical structures in string theory. This property is deducible from the fact that open strings can only end on D-branes. Indeed, an open string stretched between D-branes produces some spectrum of fields' quanta; then it produces the fields propagating along the D-branes on which the string is attached. Their dynamical nature, along with the fact that they seem to carry charges (relative to a field which is a generalization of the electromagnetic field), made D-branes essential players in the second superstring revolution and more specifically in the setting analyzed in this chapter (Polchinski, 1995).

5.2 AdS/CFT correspondence: holographic space

The original AdS/CFT correspondence was derived by combining two different descriptions of D-branes, one in terms of open strings and the other in terms of closed ones.[3] In the first case, N coincident D-branes at weak string coupling are entirely described by the massless modes of open strings attached to them. The number of D-branes is kept small to minimize the gravitational effects. And in doing so, open strings remain decoupled from the closed. In the second case, increasing the number of D-branes produces a scenario in which gravitational effects are dominant. So, the branes are entirely described by the modes of closed strings. The gravitational effects produce in turn a curved geometry. The prototypical case of such geometry is that of the $AdS_5 \times S_5$.

The two descriptions gained by varying the number N of D-branes are equivalent in virtue of the open-closed dualities of string theory established in 1989 by a paper of Polchinski, Dai, and Leigh (1989). So, the key idea that led to the AdS/CFT correspondence was that there is a decoupling regime between closed and open strings in which the low-energy theory of the open strings (Yang-Mills field theory propagating on the brane) is by itself equivalent to the low-energy limit of the theory of closed strings propagating in the throat of the bulk (Type IIB superstring theory). The dimensionless parameters of the IIB superstring theory on $AdS_5 \times S_5$ are the string coupling g and the radius $\frac{R}{\alpha'}$ of the compact dimensions S^5 – the radius is expressed in units of the string length. The dimensionless parameters of the SU(N) Yang-Mills theory instead are the coupling g_{YM} and the constant N (the number of D-branes).

The relations connecting the gravitational and the gauge parameters can be written in a quite simple way (Zwiebach, 2009, pages 541–543, also cited in Vistarini, 2016, page 128). There are mainly two relations:

$$g = \frac{1}{4\pi} g_{YM}^2 \tag{5.1}$$

and

$$\frac{R^4}{\sqrt{(\alpha')^2}} = g_{YM}^2 N. \tag{5.2}$$

Now, a third coupling constant λ of the Yang-Mills theory, also called the 't Hooft coupling, can be expressed as a function of the other two Yang-Mills parameters mentioned above, that is, $\lambda = g_{YM}^2 N$. By using this identity in (5.1) and (5.2), it is possible to rewrite both of them in the following way:

$$g = \frac{\lambda}{4\pi N} \tag{5.3}$$

and

$$\frac{R}{\sqrt{\alpha'}} = \lambda^{\frac{1}{4}}. \tag{5.4}$$

These four identities are not at the center of the formal articulation of the AdS/CFT correspondence. However, they reflect a main conceptual feature of the holographic duality.[4] The first and third identities show that both weak Yang-Mills couplings are directly proportional to the weak string coupling g. One may think that in a weak coupling regime, it is possible to make both theories simultaneously tractable, hence making the AdS/CFT testable. But the direct proportionality between R and λ in the fourth identity shows that in order to have tractability of the gauge theory, we need weak coupling and a small 't Hooft parameter λ. However it also shows that tractability of the gravitational theory can be gained if we have weak coupling, but in relation to a large λ since the curvature must be kept small.

So, the four identities describe in what sense the AdS/CFT correspondence cannot be experimentally tested: the two sides are complementary; that is, those effects making computations impossible on one side, make computations possible on the other (Zwiebach, 2009, page 543). But intractability is one side of the coin. The other side is the symmetrical inter-theoretical relation between the two sides; namely, any manipulation of the physical parameters on one side directly affects the physical parameters on the other one.

The AdS/CFT correspondence is a one-to-one map between the spectra of the two theories, with a perfect match of observables on each side. Observables in the CFT are expectation values of local operators.

The latter are mapped by the duality to observables of the string theory side. This correspondence is thought to be holographic mainly in virtue of one of its formal features, namely, the covariance energy-radius.

The basic idea is that a type of string theory describes dynamics taking place throughout the interior of the bulk. The boundary data of the duality, describing a gauge field theory, encode these bulk string dynamics. In particular, the boundary data control two types of extra dimensions in the bulk AdS_5: the bulk radius and the compact extra dimensions S_5 located at each point of the bulk. The formal articulation underlying the two cases is the same. Here for simplicity reasons, I only consider one case of extra dimensions' appearance, i.e., that of the bulk's radius.

The radial variation of the bulk corresponds in the boundary data to the variation of energy scale for the field theory. Physical events taking place in the bulk at great distance from the boundary correspond in the boundary to infrared (IR) processes, whereas events taking place at small distances from the boundary correspond in the boundary to ultraviolet (UV) processes. In other words, the radius of the bulk "is" an internal degree of freedom of gauge particles living in the boundary. To unpack this point via some main sources in the physics literature, see Vistarini (2016); in particular, the covariance relation between bulk's radius and boundary's energy scale I get here heavily relies on Boer, Verlinde, and Verlinde (2000) and Balasubramanian and Kraus (1999). To introduce a basic formulation of the covariance, we need first to introduce some terminology. Let ϕ be the bulk field solution of the gravity action with some fixed boundary values; let ϕ_{boun} denote the boundary field related to the bulk field by those fixed boundary values; finally, let's use $U(\phi)$ to denote the potential of the bulk field ϕ propagating throughout the bulk.

Now, the bulk field and the boundary field get to know each other over the boundary. A simplified version of the covariance is the following:

$$\beta(\phi_{boun}) = \frac{[\ldots]\partial U(\phi)}{U(\phi)}, \tag{5.5}$$

where on the left side $\beta(\phi_{boun})$ denotes the renormalization group flow of the boundary field – that is, the beta functions describing the change of the coupling of the boundary field with respect to the change of the energy scale happening over the boundary. The right side contains information about the radial evolution of the bulk field coupling with closed strings and other bulk fields; the quantity within the square parenthesis (omitted for simplicity reasons) refers to the couplings of the bulk field.

Although it is not the canonical equation of the AdS/CFT correspondence, the formula (5.5) illustrates in simple formal terms why the AdS/CFT is called holographic. The formula also points to the symmetrical structure of such correspondence by completing the picture delivered by the four identities (5.1)–(5.4). Indeed, (5.5) shows that the renormalization group (RG) flow of the boundary field theory is in a one-to-one correspondence with the variation of the potential of the scalar field radially propagating in the bulk (Vistarini, 2016). Then, what one gains here is some dictionary translating the formal articulation of the boundary gauge field theory into that of the bulk string theory (and vice versa). The physical content does not get lost in translation though. Indeed, the two theories have the same physical content in virtue of being holographically dual.

Then, it is exactly for the existence of this physical equivalence between the two sides that there is no room for any talk about emergence of the bulk extra dimensions from the boundary. As I will argue in the second half of this chapter, the AdS/CFT correspondence delivers a notion of emergent space. But not how one might expect. Indeed, given the formal features and the physical content of the correspondence, the spatial extra dimensions in the bulk are emergent in the same way as the corresponding dual data on the boundary are. Spacetime emergence in the holographic duality involves both sides. The formal structure of the energy-radius covariance does not support the idea that one of the two sides of the duality may be emerging from the other – see also Vistarini (2016).

5.3 Deconstructing spacetime emergence in holography

As I said, the boundary and bulk theories are different. What "different" means in this context is a crucial philosophical topic unpacked in this section, since the notion of difference to which it refers is crucial to understand spacetime emergence via AdS/CFT duality. Let's start with saying that one feature of this difference is that the two dual theories posit different geometries (and a different number of dimensions), whereas a second feature is that the two dual theories also posit different physical ontologies. This point will be developed in the first subsection.

In spite of the difference, the two theories share the same exact physical content, that is, they make precisely the same predictions about all observable phenomena. As we saw, the holographic duality (and dualities in general) relates *via* a one-to-one map any physical property of any system of the theory on one side with that of the theory on the other. The expected values of any physical property in any state are the same in both theories, despite the fact that the "types" of physical properties and of physical systems involved in both theories are not the same. As we saw in the previous section, the bulk's radius seems to be encoded in the boundary as an internal degree of freedom of the gauge particles. More precisely, the radial extra dimension appears "to be the same as" the RG flow of the boundary (which does not contain that spatial dimension).

Now, within the quantum gravity circles, the philosophical debate on spacetime emergence sometimes uses holographic duality as some sort of grounding relation. More precisely, one side of the correspondence, usually the bulk's radius, is thought to emerge from the dual gauge theory.[5] Elsewhere, I argue against this way of reading holographic duality. This duality cannot be interpreted as a form of grounding relation, unless it is possible to prove some asymmetry between the physical contents of the two theories. A way of doing so might be that of showing that the boundary theory is characterized by a lower number of physical degrees of freedom than that of the bulk. The lower number of spatial dimensions of the boundary cannot be considered to be indicative of the fact that the boundary has less physical degrees of freedom than the bulk has. Indeed, it is not indicative because *in primis* there isn't any direct relation between the number of physical degrees of freedom of system dynamics and the number of spatial dimensions in which those dynamics unfold and last but not least, because – as far as we saw in the previous section – the physical degrees of freedom on each side actually perfectly match via the one-to-one correspondence. For this reason, to say that the bulk radius "emerges" from the boundary dynamics sounds like the wrong way of introducing spacetime emergence within the context of holographic duality. The bridge-like features of the formal articulation of the AdS/CFT duality, and its physical content, do not account for the conceptual asymmetry built into the notion of emergence (Vistarini, 2016). Similar argumentative lines are developed by Teh (2013), Rickles (2012), and Dieks, van Dongen, and de Haro (2015). In this section, I am deepening the argumentative line presented in Vistarini (2016).

Now, there are at least three topics that need to be deepened. One amounts to be an explanation of why I claim that the relation of emergence is naturally provided with asymmetric features. Another related topic concerns an attempt of understanding what about the AdS/CFT duality is incompatible with the idea of an AdS bulk emerging from the boundary and finally, a proposal of explanation of where exactly in the context of holographic duality one might trace spacetime emergence. At the end of the day, one might say that holographic duality, as T-duality and string mirror symmetries, shows a criterion of non-fundamentality, rather than identifying directly and exhaustively what is fundamental in string theory's physical ontology.

5.3.1 *Holographic space: emergent in what sense?*

As I said, holographic duality defines a criterion of non-fundamentality. Either side of the duality is not more fundamental than the other because of their exact physical equivalence. The AdS theory on one side is physically equivalent to the CFT on the other. However, they look different.

Trying to explain in what sense they differ, one might say that, for example, an observable in the quantum field theory side is the expectation value of some local operators. This observable gets mapped onto an observable of the gravitational side that amounts to be the perturbation of some boundary condition. So, some values of some local physical property in a world without gravity is mapped onto some physical interaction happening in a world with gravity: the mapping is between different dynamical structures described by different mathematical languages.

In addition to this formal incommensurability, the two theories posit different spacetimes. The difference in this case can be found in the fact that it would take a discontinuous and abrupt deformation to get one geometry from the other. Finally, there is a broader difference between the two theories, one that is unveiled by the incomparability of their physical ontologies. One might say that each description of what sort of physical reality populates the fundamental structure of the world is ontologically irreducible to the other. This notion of irreducibility will play a central role in defining one of the two meanings in which spacetime emergence shows up in the holographic duality.

Now, in order to assign a role to spacetime in the picture of holographic duality, I need to shift the attention from the physical content of the two theories, which they share exactly, to the set of those theoretical features neglected by the holographic translation. What the holographic duality (and all dualities in general) does not detect is what is not fundamental in the two theories' physical ontologies – see also Polchinski (2015) and Vistarini (2016). Spacetime is exactly in this set of neglected features, along with other components of the theories' ontologies. For example, on one side of the holographic duality, the ontologically primitive objects are fields, whereas on the other, they are strings. Note well that these differences do not entail any degree of distinction between the theories' physical contents. Equivalently, those differences do not produce a notion of irreducibility involving the physical contents. This holds true in virtue of the duality's dictionary that establishes the physical reduction in two ways (i.e., physical equivalence).

Now, that said, I claim that there are two coexisting meanings according to which emergent spacetime fits the holographic picture. The first is transversal to this picture, and it enhances the philosophical thickness of this duality by introducing the emergent nature of the duality itself. Let's unpack this point. The diversity between the theories of the dual pair is a difference regarding the ways in which the shared physical content gets represented at some less fundamental level of physical reality. This physical content is thought to be emergent itself, to be arising via supervenience from some more fundamental dynamical degrees of freedom. So, geometry, number of dimensions, ontologically different primitive objects (strings on one side, fields on the other), and so on are features of the way in which the relevant physical information is represented at some higher level. Then, in this sense, the two spacetimes each appearing on each side of the holographic duality are different patterns of representation, equally emergent in virtue of the emergent nature of the AdS/CFT correspondence[6] (Vistarini, 2016).

Every physical state on each side of the duality might be thought as supervening some underlying non-spatiotemporal relation structuring some distribution of fundamental units. Achieving a topological formulation of this presumed non-spatiotemporal relation might count as a way of identifying the presumed underlying dynamics. Moreover, in the AdS/CFT scenario, we don't have only sets of physical states on each side. We also have a relation of covariance among them. My interpretation of this covariance (which unveils a double aspect physical scenario) is in terms of a symmetrical correspondence supervening some deeper covariance, some non-canonical relation of "quantum entanglement" between sets of physical states, each living in different spacetimes. The latter would be "connected" by some type of discontinuous, abrupt deformations, not preserving their topological structures, yet preserving the two theories' physical content.

Here, "entanglement" might not be the best characterization for this covariance, although the non-locality of traditional entanglement in quantum mechanics undermines the fundamentality of spatiotemporal relations as well. However, traditional quantum entangled states share the same spacetime, whereas here the covariance takes place between pairs of physical states in two different spacetimes. So, by attempting to conceive an extension of ordinary entanglement, the "entanglement" might work like a sort of thread stitching the two spacetimes together.

The second meaning of spacetime emergence is internal to the holographic picture. What I mean with "internal" is unpacked in what follows. From the point of view of each theory of the holographic dual pair, the physical content of the other does not appear emergent in virtue of the duality translation establishing physical reduction in two ways. However, as I said above, holographic duality does not detect geometry. In other words, in the duality dictionary, there isn't any explicit formula translating spacetime structure on one side into some different spacetime structure on the other side. One might formulate here a definition of *irreducibility* not involving the physical content of the two theories, rather involving whatever lies outside that physical content: structures of either theory that do not get translated by the duality in structures of the other are the irreducible properties. This irreducibility is what produces the overall diversity of the two theories without breaking their physical equivalence (Vistarini, 2016).

Now, the geometry of the bulk and that of the boundary do not get translated by the holographic duality. So, from the perspective internal to the boundary theory, the overall AdS spacetime is irreducible to anything in that perspective. Therefore, it is emergent because it appears like some feature coming out from nowhere. And the same holds in the other way around. From the bulk theory's perspective, the boundary spacetime structure is emergent, because it does not reduce to anything in the bulk description. In other words, this second notion of spacetime emergence is internal to the theoretical perspective chosen. It develops around a notion of irreducibility dictated by the theoretical perspective taken in consideration (Vistarini, 2016).

5.3.2 *Why supervenience-based emergence of the bulk from the dual boundary doesn't work*

The standard philosophical formulation of supervenience-based emergence is that delivered by McLaughlin (1997):

> If P is a property of w, then P is emergent if and only if the following two conditions are met: (1) P supervenes with nomological necessity, but not with logical necessity, on properties the parts of w have taken separately or in other combinations; (2) some of the supervenience principles linking properties of the parts of w with w's having P are fundamental laws."[7]

Now, the second condition unambiguously shows why emergence *via* supervenience does not work when applied to the relation between the two sides of the holographic duality. Indeed, that part says that in order to establish if a physical property is emergent, one first needs to have an independent criterion of fundamentality. That shows how an asymmetry between degrees of fundamentality is built into this type of emergence since the start. And that is the main point making this notion inapplicable to the relation between the two sides of holography.

This characterization of emergence has been already applied to general relativistic spacetime emergence. That is not surprising since McLaughlin's definition naturally applies to all those cases in which the emergent laws arise *via* an asymmetrical physical change of energy scale – asymmetrical in the sense that the change is not a duality and that some physical content is lost in translation. This asymmetrical change provides an independent criterion of fundamentality. The derivation of general relativity laws by reimposing conformal invariance of the quantum string laws necessarily requires this asymmetrical change of energy scale. For this reason, the derivation gains an asymmetric formal structure necessary to the grounding – see Vistarini (2016).

The formal asymmetry of the derivation, along with the bottom-up metaphysical reading, provides a criterion of fundamentality establishing when it is the case that a property is emergent. Note that the derivation of general relativity from quantum string theory does not involve duality. The two theories are not physically equivalent. Now, in the case of holographic duality, nothing in its formal structure shows some trace of asymmetry. One element of the dual pair can be derived from the other and vice versa. Nothing in the formal structure of the energy-radius covariance shows uneven levels of fundamentality. Therefore, that is why supervenience-based emergence cannot appropriately describe the inter-theoretical relation of the dual holographic theories.

5.3.3 Why traditional epistemic emergence of the bulk theory from the boundary one does not work

As I said in Chapter 1, within the spectrum of views defining emergence in epistemological terms, a main school of thoughts is that defining emergence in terms of irreducible patterns.[8] What does irreducible pattern mean in the traditional debate on epistemic emergence? As I said in Chapter 1, usually an irreducible pattern is a property of some system ruled by some law-like generalization within the domain of special science (Fodor, 1974; O'Connor and Wong, 2015). The property in question can be identified only within the higher level of description of the system delivered by special science. In this sense, such property is irreducible to other properties populating a lower (more fundamental) level of description of the system. Irreducibility here is due to pragmatic reasons. The attempt of deriving principles of psychology directly from principles of quantum mechanics inevitably must face an almost insurmountable obstacle. Even if possible in principle, explaining psychological states by appealing to some quantum Hamiltonian turns out to be de facto non-achievable because of computational obstacles. Here, the asymmetric structure of the formal relation between psychology and quantum mechanics is supported by the huge difference of their respective subject matters. A special science as psychology has a scientific content which is irreducible to that of a physical theory like quantum mechanics, because properties and law-like generalizations described by the scientific content of the former cannot be entirely captured by physical content of the latter.

Now, one may try to read traditional epistemic irreducibility into the relation between physical theories whose physical contents apply to comparable physical length scales in space – that is, their respective physical parameters refer to domains of application close enough. The AdS bulk theory and the gauge boundary dual partner may exemplify a case like that. Then they might be seen as two irreducible theories in the sense of traditional epistemic irreducibility, both living at comparable spatial length scales (Vistarini, 2016). But this is not going to work. Indeed, the main core of their holographic relation is an unambiguous formal bridge connecting them. It is in virtue of this bridge that dynamical patterns on one side completely capture dynamical patterns on the other side. The theories' physical contents are not lost in translation, not even partially (Vistarini, 2016).

That explains why traditional epistemic irreducibility does not capture this peculiar duality relation. The former delivers a notion of irreducibility that applies to the theories' physical contents, and if applied to two physical theories about the same length scale, it would necessarily produce a scenario in which the theories would become competitors. But dual physical theories cannot be competitors since they share the exact same physical content.

Notes

1 Both the AdS/CFT duality and the dS/CFT conjecture involve non-perturbative string theory. While the AdS/CFT keeps time out of the story (the emergence is purely spatial), the dS/CFT conjecture involves time. The conjecture is the only string theory context in which time gets an exclusive treatment, according to which time acquires the presumed emergent status of a hologram (Vistarini, 2016).
2 Be reminded that, as we saw in Chapter 3, at low energy, quantum string theory's dynamical laws "are" the general relativity ones; hence, one can meaningfully say that the AdS spacetime (without compactification) is already a solution of quantum string theory via the low-energy derivation analyzed in Chapter 3.
3 This section heavily relies on Vistarini (2016) and on the main bibliographic references used in that paper for the technical parts — in particular, Maldacena (1998), Polchinski (1995), and Zwiebach (2009); also Balasubramanian and Kraus (1999) for the description of how some boundary dynamics control the holographic radius of the bulk; and finally, Boer, Verlinde, and Verlinde (2000).
4 Technical detail about how exactly the physical parameters on each side map onto each other can be also found in Polchinski (2015, page 30).
5 For the sake of accuracy, not only the bulk radius but also the compact extra dimensions at each point of the bulk are considered to be emergent structures.
6 Note well that this line of reasoning is not in tension with the fact that both the AdS and Minkowski spacetimes involved in the AdS/CFT correspondence are physical solutions of general relativity. All I am saying here is that, although they are real in virtue of being physical solutions of an empirically robust low-energy theory, they are byproducts of more fundamental dynamics.
7 Ibid. Also see O'Connor and Wong (2015) and Vistarini (2016).
8 For more detail on epistemological emergence, see Chapter 1, O'Connor and Wong (2015), and Vistarini (2016).

6 String theory's background independence explained through the theory's moduli space

6.1 Sketch of the view pursued

In this chapter, I try to read string theory background independence through the local structure of its "moduli space," and in so doing, I present an argumentative line unifying those separately developed in the chapter on T-duality and in that of the anti-de Sitter (AdS)/conformal field theory (CFT) correspondence. The argumentative line also latches on that initiated in Chapter 1 to argue that spacetime emerges from underlying quantum strings laws, rather than being a fixed arena somehow prior to those laws. Let me summarize here the main ideas presented on this topic in Chapter 1.

At first, I considered some classical Hamiltonian on an actual Euclidean flat three-dimensional space. Such Hamiltonian gives rise to Euclidean flat three-dimensional appearances. Then, I showed that the same Hamiltonian on an actual arbitrarily curved space produces Euclidean flat three-dimensional appearances as well. So, I concluded that the manifest geometry of the world is produced by the Hamiltonian regardless of the actual curvature of the space. But things appeared to be unequivocally more radical than that: the dynamics produce a manifest geometry of the world whether or not there is a fundamental geometry at all. Then, the fundamental structure of the world might well be something weaker than a geometry, something like a fundamental (non-geometrical) topology.[1] Thinking of this possibility, the necessary condition for producing flat Euclidean appearances might be something like a distribution of matter having some topological relational structure such that a coordinate system on that structure exists where the distribution is governed by a Hamiltonian producing these geometrical appearances.[2] Loosely speaking, the topology of some matter distribution describes how the elementary parts of that distribution relate to each other. In other words, a claim about the topology of some matter distribution is a claim about the dynamical interactions constituting that distribution and obeying some set of laws. Then, a claim about that topology might be seen as a claim about some set of fundamental laws.

However, when it is about string theory, things might get complicated: a claim about the topology of some distribution in some cases might not be enough to identify a claim about some set of fundamental laws. So far, we discussed the necessary condition for producing the manifest geometry of the world. What about the sufficient condition? Now, the sufficient condition for producing flat Euclidean appearances would be obtained only by identifying explicitly the topology in virtue of which some coordinate system exists such that the Hamiltonian in it would produce these appearances. But the thing is that in string theory, the search of a sufficient condition for producing the manifest image of the world might not identify a unique topology because of the existence of mirror symmetries. These string dualities preserve physical content, but some of them do not preserve topological features.[3] As I said in Chapter 1, back in the mid-1980s according to quantum string theory, there were only few different topological shapes compatible with low-energy-physics matching phenomena. So, people assumed that, if the fundamental structure of the world has a unique topology, one can try to analyze each, searching for a match to known physics. Years later, the number of topological shapes compatible with low-energy physics had grown to a few thousand, continuing to multiply until the late 1990s. Looking for a mathematical directive from the theory that might single out a particular shape as the right one was an impossible task.

Now, the under-determination of topology by string physics shown by some string mirror symmetries can be read in the same way in which I read in Chapters 1 and 3 the under-determination of geometry by string physics: trying to single out a particular shape as the right one might be the answer to the wrong question. The right problem might not be about finding the right geometry or topology inhabiting the fundamental structure of the world, because this problem assumes there is one. Perhaps the mathematical directive from quantum string theory is that of looking for some more fundamental relational structure, something like sets of deformations of geometries and also of topologies. The set might range from continuous deformations changing geometry but preserving topology to increasingly discontinuous ones via which topological invariants are not preserved. In this sense, repeating the point that a claim about topology might be identified to a claim about dynamical laws, one might argue that some "law-like" deformations of laws might be more fundamental than laws. This is what we are going to explore.

6.1.1 Familiarizing with moduli spaces

As I said in Chapter 3, the set of data necessary to define string theory includes the notion of moduli space.[4] The latter is a "space" that can be used as some kind of navigation system through the physical content of the theory, hence allowing an indirect exploration of those physical features not directly accessible by experiments. A moduli space is not a physical spacetime; rather, it is an abstract space whose local structure – in this chapter – encodes information about families of physical spacetimes, along with the physical parameters of physical theories over those spacetime. In general, moduli spaces are used for a variety of goals. As I said in Chapter 3, they were imported from algebraic geometry and have been originally used to parameterize families of mathematical structures. Once imported into theoretical physics, they have been also used to parameterize any sort of physical structure.

Now, to complete the basic identikit shared by any moduli space that has been initiated in Chapter 3, a simple characterization might be the following: a moduli space M is a *space of parameters* for some family K of objects *only if* the correspondence between "points" of M and objects in K is *well defined*. In this context, the expression *well defined* is used in a strict mathematical sense. That is, given a function ϕ, namely,

$$K \downarrow^{\phi} M$$

ϕ is a well-defined function just in case each object in K is mapped on a unique point of M. This feature does not need to be satisfied in the other way around. That is, in order for ϕ to be *well defined*, its inverse ϕ^{-1} does not need to be *well defined*. Equivalently, the set $Ker(\phi)$ is not required to be equal to the empty set $\emptyset$.

Now, this requirement by itself would only produce some humble structure. Some more constraints should be added in order to have something richer and useful for applications. Then, we might require also that whenever two objects in K are somehow "similar" or "close" with respect to some property, the corresponding "points" on M must be close as well with respect to some topology on the moduli space M – as we will see, in my specific case, I will define this topology in terms of a form of similarity order. This second constraint induces a refinement of the basic property of being well defined: things in K that get mapped onto the same point in M must be at least quite "similar" or "close."

Now, these two requirements combined together produce a simplified version of a "fine" moduli space for some sets of objects. This structure, once enriched with additional formal features, turns out to encode as much rich information as possible about those objects. Indeed, it possibly encodes intra-relations among parts of a single object and also inter-object relations.[5]

Next, how does string theory's moduli space relate to an argument for background independence? Answering the question requires focusing on the specific profile of the string theory moduli space I am drawing in this chapter. A main philosophical goal is that of showing (via this space) the genuine metaphysical commitment of the theory to spacetime non-fundamentality. It will eventually turn out that some fiber bundle structures (conveying physical information) on top of the moduli space's topology, make explicit this commitment.

Now, as I said in Chapters 1 and 3, string theory physics circles use the theory moduli space with methodologies and purposes that do not overlap with the ones I present in this chapter. On the one side, this space is often used for classification purposes, that is, to complete the duality networks connecting different formulations of the theory and to show that either different formulations are trivially the same or they can be obtained from the same more complete formulation. The hope here is that of finding the universal formulation from which any other existing formulation can be gained by pullback. On the other side, this space is used to explore many cosmological applications of quantum string theory. As I said, the moduli space I build works like a sort of unusual global positioning system. What one identifies by analyzing its local structure has nothing to do with ordinary geographical positions in space; rather, it has to do with the degree of independence of string physics from spacetime.

6.2 Spacetime metaphysics within the string physics

One may say that string theory moduli space represents the *totality of possibilities* for stringy worlds to be in a way or another. Somehow this totality is "fundamental" in two senses. First, it shows unambiguously what in the theory is not fundamental: geometry. T-duality and the AdS/CFT duality show that both ordinary spacetime and compact extra dimensions are emergent in the theory. Second, it shows what might be more fundamental than geometry and in some cases than topology: their deformations.

For the sake of simplicity, let's import here the same basic formal notations of Chapter 3. As we saw, the higher-dimensional spacetimes in the theory are formally denoted by product spaces of the type $S \times K$, where S is usually picked from the sets of general relativity solutions, and K denotes any set of compact extra dimensions allowed in the theory.

As we saw, any K in the theory can be viewed as a compact manifold, and the word "manifold" here simply denotes any topological and geometrical structure we are allowed to consider (there is no underlying substantivalist view). Note well that, as a matter of fact, some of the topological invariants of these structures, namely, invariants with respect to transformations that change their geometries, are responsible for yielding observable physical phenomena occurring in S.

Now, what are exactly these transformations changing the geometry of any K that can be found in the theory? In this chapter, I'll be using some basic techniques of deformation theory. Deforming a geometrical structure can produce a great variety of geometrically different ones, and it can be done in many different ways. Here, I am restricted to smooth deformations.[6] Although any class of deformations can apply to any part of the background $S \times K$, here for simplicity reasons, I am considering only smooth deformations applied only to the compact part K.

Smoothly deforming any K produces some family of structures more or less similar to K. The family obtained surrounds the original copy, but the notion of "surrounding" here does not evoke an ordinary notion of spatial closeness. Being close on top of the moduli space means having a high degree of "similarity." Note well that one may produce any family of deformations without any prejudice to choice of a particular member as the starting point of the deformation process. So, any arbitrarily chosen starting point will be called the central fiber of the family, and it will be denoted by K_0. Instead, any K_λ will be denoting any smooth deformation arising from the process, and it will be also called the generic fiber.

Now, by gluing together fibers (and families of fibers), we build the moduli space topology. That is true of any moduli space. Then, on top of this topology, we can put some extra structures that keep track of dynamics of some physical system. The information recorded by local areas of the moduli space depicts the dynamics of a physical system against different spacetimes, each spacetime being a fiber of the family and each spacetime obtained as deformation of another spacetime. In our case, the fiber bundle structure will keep track of some string system. In this way, we can construct a string theory by starting from data that amount to be families of spacetimes.

Now, a first characterization of background independence may already be formulated at this stage in simple terms. Broadly speaking, what one wants to show in this case is that a set of families of geometries (spacetimes) can be taken as the data for constructing a string theory without any prejudice to choice of a particular family. If that can be shown, then the theory is background independent. But to fully understand this provisional formulation, one needs to say something about the dynamical aspects of this formal construction. Leaving aside for now any consideration about the dynamical structure that will be posited on top of the moduli space, let's focus here on some more precise characterizations of the moduli space topology.[7]

What about the gluing conditions keeping together the local patches of the theory's moduli space? This is an entirely topological point revealing the formal connection between physical spacetime geometry and the topological structure of the moduli space. As I said, K_0, along with the other family members K_λ, are formally represented as compact manifolds. In this chapter, for reasons connected to the canonical mathematical literature on the topic, these compact manifolds are also complex.

What kind of mathematical object is a complex compact manifold? A *complex manifold* is a topological structure that can be entirely covered by an atlas of charts, each patch representing a unit disk in the ordinary numerical space $\mathbb{C}^n$. Each patch overlaps with its neighbors in those regions in which they are stitched together. So, we need to make sure that for any given manifold covered by an atlas, the values of any local function at any point do not depend on the variation of patches sharing that point. In other words, we also need to introduce transition maps from any patch to any other one in its surrounding. In order to guarantee everywhere consistency and smoothness of the local stitching, *holomorphic* transition maps are used to glue together different patches in the same neighborhood.[8]

6.2.1 *Smoothly deforming compact manifolds*

Let's dig into the notion of family of deformations. The idea we can start with is that of a map associating an arbitrary parameter λ to a complex manifold K_λ. A map like that behaves pretty much in the same way as a traditional function of λ would do. We can think of that map as a function defined over a set of variable parameters λ that takes some value on a fixed parameter λ_i and some other value on a different fixed parameter λ_j. Each "value" taken by the "function" is a complex compact manifold K_λ. So, unlike a genuine function, there isn't a unique space containing the whole range of function's values.[9] Lets' unpack this point. I introduce a new structure K defined in the following identity:

$$K = \bigcup_{\lambda \in B} K_\lambda,$$

where B is a numerical domain contained in the numerical field of complex numbers $\mathbb{C}$. The choice of a one-dimensional space of parameters is due to simplicity reasons.[10]

We will be better off if K_λ are not simply values taken by a "function" of λ but rather by a C^∞ "function" of λ. By imposing this infinitely smooth functional dependence, we can easily obtain differentiable families of compact complex manifolds.[11] Being a differentiable manifold simply means being equipped everywhere with a local structure completely resembling a linear space, that is, a place where we can use familiar calculus.

The following definition describes the sufficient conditions to have a differentiable family of compact manifolds:

Definition 6.2.1. Suppose we are given a domain B in $\mathbb{C}$ and a set $\{K_\lambda | \lambda \in B\}$ of complex manifolds K_λ depending on λ. We can say that K_λ will have a C^∞ dependence on λ and that $\{K_\lambda | \lambda \in B\}$ is a *differentiable family* of compact complex manifolds, if there are a differentiable manifold K as above and a C^∞ map ϕ of K onto B satisfying the following conditions:

1 The differential of ϕ, $\phi_\star: T_pK \longrightarrow T_{\phi(p)}B$ is surjective at every point $p \in K$, where T_pK and $T_{\phi(p)}B$ are, respectively, the tangent space to K at the point p and the tangent space to B at the point $\phi(p)$.

2 K is a non-empty complex compact manifold and for each $\lambda \in B$, $\phi^{-1}(\lambda) = K_\lambda$ is a compact differentiable submanifold of K.

3 There are locally finite open covering $\{V_j | j = 1,2,\ldots\}$ of K and complex-valued C^∞ functions $z_j^1(p),\ldots,z_j^n(p)$, $j = 1,2,\ldots$, defined on V_j such that for each λ, the following coordinatization forms a system of local complex coordinates of K_λ[12]:

$$\{p \to (z_j^1(p),\ldots,z_j^n(p)) | V_j \bigcap \phi^{-1}(\lambda) \neq \emptyset\}.$$

So, λ is the parameter of the differentiable family $\{K_\lambda | \lambda \in B\}$, and B is the family parameter space or its base space.

Now, the coordinate transformation defined on $V_j \bigcap V_k \neq \emptyset$,

$$z_k(p) \to z_j(p),$$

can be rewritten as functionally dependent on λ:

$$(z_k^1(p),\ldots,z_k^n(p),\lambda) \to (z_j^1(p),\ldots,z_j^n(p),\lambda)$$

$$= (f_{jk}^1(z_k(p,\lambda),\lambda),\ldots,f_{jk}^n(z_k(p,\lambda),\lambda),\lambda),$$

where $f_{jk}(z_k,\lambda)$ on $V_j \bigcap V_k \bigcap \phi^{-1}(\lambda)$ are C^∞ functions of λ, and also they are holomorphic functions of z_k.

These functions are very important since they dictate how the open neighborhoods V_j and V_k join together for every λ. Their explicit functional dependence on the variable parameter λ shows their central role in deforming a complex compact manifold: deforming the gluing functions of the manifold, one produces another one more or less similar to the original, depending on the type of deformation.

So, we now have a notion of a differentiable family (K,B,ϕ) of complex compact manifolds. But here we need a specific type of differentiable family of complex compact manifolds. We are looking for a family of *deformations* of some central element K_0. Namely, given a compact complex manifold K_0, if there is a differentiable family (K,B,ϕ) with the features described in the definition above and in which $\phi^{-1}(0)$ is exactly K_0, then each $K_\lambda = \phi^{-1}(\lambda)$ is a deformation of K_0. One may also call K the total space of deformations of K_0.

In other words, deforming in some way or another the compact manifold K_0 affects how the open sets $V_1, \ldots, V_j$ glue together. That is, it affects the functional dependence on λ of the coordinate transformations $f_{jk}(z_k, \lambda)$.[13]

Deforming the gluing functions $f_{jk}(z_k, \lambda)$ of a compact complex manifold means deforming its complex structure. Although all the K_λs of the family, along with K_0, share the same differentiable structure, each of them carries its own distinct complex structure. Generally, a complex structure over a manifold (whether or not compact) comes along with a Hermitian metric compatible with it, which in principle is not unique.[14]

There is no a priori fixed relation between metric (or geometry) and complex structure; rather, there are many different compatibility conditions, namely, many different mathematical correspondences involving holomorphic functions in the manifold's atlas of charts. I won't give here detail on such mathematical relations. It suffices to say that given a compatibility condition between complex structure and an induced metric, any deformation of the complex structure is also a deformation of the geometrical one.

Therefore, we have a family of geometrically inequivalent, diffeomorphic backgrounds. Note well here that the differentiable structure of any individual spacetime in the family is not the same as the "differentiable" structure of the moduli space. The latter is a topological structure technically arising from the C^∞ deformations of one spacetime into another. Within this type of family of deformations, we now pick a specific subtype, namely, first-order infinitesimal deformations.

6.2.2 *First-order infinitesimal deformation*

Let's consider $\lambda = 0 \in B$ such that $\phi^{-1}(0) = K_0$, namely, the central fiber of the deformation family. I now confine my attention to an infinitesimal interval containing 0 or equivalently to a first-order neighborhood B_ϵ of 0, i.e.,

$$B_\epsilon = \{\lambda \in B | \lambda \in (-\epsilon, \epsilon)\}.$$

If one has a differentiable function over this interval, one can know its infinitesimal variation around 0, by differentiating the function in λ.

As we saw above, our differentiable family differs in many aspects from a differentiable function. Nevertheless, we can similarly perform the differentiation in $\lambda \in (-\epsilon, \epsilon)$ of

$$f_{ik}(z_k, \lambda) = f_{ij}(f_{jk}(z_k, \lambda), \lambda), \tag{6.1}$$

over $V_i \cap V_j \cap V_k \neq \emptyset$.

A more rigorous style of notation should also take into account that each transition function has a finite number of components. Indeed, for the sake of precision, they should appear with one more index, namely, as $f_{ik}^{\alpha}(z_k, \lambda)$ indexed by the coordinates $\alpha = 1, \ldots, n$. However, sticking to more rigorous notations would not help to understand the logic underlying this formal treatment. Here, for our purposes, it suffices to use the simpler formal notation $f_{ik}(z_k, \lambda)$.

Taking the derivative in λ on both sides of (5.1), we have

$$\frac{\partial f_{ik}}{\partial \lambda} = \frac{\partial f_{ij}}{\partial f_{jk}} \frac{\partial f_{jk}}{\partial \lambda}. \tag{6.2}$$

But why am I taking the derivative and not some other formal manipulation of the transition functions (or gluing functions)? The final goal is that of discovering what happens to the physics of some string system originally "embedded" in a certain spacetime, if such spacetime is infinitesimally deformed. The derivative of the gluing functions above gives the rate of their change with respect to the parameter λ of the family. In this sense, the derivative gives information on all the directions along which the original spacetime may bend, if deformed. Each bending yields a geometrical structure slightly different from the original.

The whole story may be retold in much more simple terms. If you glue together the pieces of some broken object changing just a tiny bit their original mutual dispositions, you get a different object that looks like the original one. If you make several attempts to glue together its pieces, each time assembling them differently from the time before and from their original mutual disposition, then each time you will get a slightly different object. This is pretty much what happens to an infinitesimal deformed geometrical structure, except that smooth deformations do not break anything.

Now, we need to express clearly all these bending activities, namely, we want to identify the appropriate mathematical language to express them. As I said, the rationale behind is the need of understanding what happens to the values of the physical observables of a string system if we change the spacetime provisionally posited as background for its dynamics. In case of background spacetime independence, geometrical change should not affect the physical observables of the system.

Now, back to an arbitrarily small neighborhood B_ϵ of 0, the act of taking the derivative of any gluing function over this infinitesimal disk produces, morally speaking, a holomorphic field at $\lambda = 0$, namely,

$$\theta_{ik}(0) = \frac{\partial f_{ik}(z_k, \lambda)}{\partial \lambda}|_{\lambda=0}. \tag{6.3}$$

Skipping any technical detail (contained in the main bibliographic source cited in endnote) by rewriting everything in θ fields and by also using differentiation rules for composite functions, the identity (5.2) gets transformed in the following one:

$$\theta_{ik}(0) = \theta_{ij}(0) + \theta_{jk}(0). \tag{6.4}$$

The latter holds true on small neighborhoods like $V_i \bigcap V_j \bigcap V_k \neq \emptyset$ – which are also open sets over K_0. The identity can also be written as

$$\theta_{jk}(0) - \theta_{ik}(0) + \theta_{ij}(0) = 0. \tag{6.5}$$

Replacing $i = k$, namely, considering only the $V_j \bigcap V_k$ part of the intersection, we get $\theta_{kk}(0) = 0$; hence, the identity (6.5) becomes

$$\theta_{kj}(0) = -\theta_{jk}(0). \tag{6.6}$$

Identities as (6.4), (6.5) and (6.6) hold true for every $\lambda \in B$. Then, let's focus on the central fiber of the family $K_0 = \phi^{-1}(0)$ and its first-order infinitesimal neighborhood parameterized by $B_\epsilon = (-\epsilon, \epsilon)$, namely,

$$\bigcup_{\lambda \in (-\epsilon,\epsilon)} K_\lambda = \phi^{-1}(B_\epsilon).$$

From (6.5) and (6.6), we can infer an important formal property of the field $\theta_{jk}(0)$, namely, that of being identifiable to a mathematical object carrying crucial dynamical information. In mathematical terms, one may say that $\theta_{jk}(0)$ is a 1-cocycle of the sheaf T_{K_0}, namely, the sheaf of holomorphic vector fields over K_0. Its cohomology class $\theta(0)$ is an element of the cohomology group $H^1(K_0, T_{K_0})$.[15] But what does it mean? Informally speaking, the cohomology group $H^1(K_0, T_{K_0})$ is simply a vector space, whose vectors (cohomology classes) can be read in more familiar terms as closed and non-exact differential 1-form.[16]

As we saw above, $\theta(0)$ represents the "derivative" at $\lambda = 0$ of the complex structure of K_λ. This derivative gives information about how space-time geometry changes with respect to the λ parameter's variation, that is, with respect to the variation of picks from the family

$$\bigcup_{\lambda \in (-\epsilon,\epsilon)} K_\lambda = \phi^{-1}(B_\epsilon).$$

The derivative is obtained differentiating in $\lambda \in (-\epsilon, \epsilon)$. In other words, we just found a vector field, namely, $\theta(0)$, bending infinitesimally the manifold K_0 along some direction, and in this sense $\theta(0)$ is an *infinitesimal first-order deformation* of K_0 along some direction

$$\frac{dK_\lambda}{d\lambda}|_{\lambda=0} = \theta(0). \tag{6.7}$$

Therefore, the cohomology group $H^1(K_0, T_{K_0})$, generated as vector space by a basis of non-isomorphic θs, represents all the infinitesimal deformations of K_0.

Now, let's conclude this section with some technical remarks preparing the ground for the next section. I just introduced the formal expression for any first-order family of deformations (equipped with the properties described above). The expression is the following:

$$\phi : \bigcup_{\epsilon} K_{\epsilon} \longrightarrow B_{\epsilon}.$$

Any family like that, that is, any family characterized by the map ϕ, always comes with another map associated to ϕ, namely, the pullback map $\phi^{\star}$. The latter looks like the following:

$$\phi^{\star} : Fun(B_{\epsilon}) \longrightarrow Fun(\bigcup_{\epsilon} K_{\epsilon}).$$

Once $0 \in B$ is fixed, the first-order neighborhood B_{ϵ} turns out to be a topological space whose sets of functions are those not vanishing at 0 and whose first-order derivatives do not vanish at 0.[17]

So, it turns out that each of these functions corresponds to a tangent vector to B at 0. Then, $Fun(B_{\epsilon})$ is the tangent space to B at 0, i.e., $T_{B,0}$.

Now, a tangent vector in $T_{B,0}$ is pulled back by the map $\phi^{\star}$ to a function over the infinitesimal family. The latter will be a function not vanishing on K_0 and whose first derivative does not vanish on K_0. So, these pullbacks describe directions on K which are normal to K_0 and along which the functions vary. Therefore, they can be thought as first-order deformations of K_0, and so they can be represented as elements of $H^1(K_0, T_{K_0})$.

6.2.3 *Kodaira-Spencer map*

The Kodaira-Spencer map has a crucial role in the context of infinitesimal deformations of compact geometrical structures. In this chapter, the map also contributes to complete the picture of how the local topology of the moduli space is actually made of "local facts" about geometrical deformations of some arbitrarily chosen physical geometry. As I said, the facts encoded by the moduli space are local not in the ordinary sense in which things are close in spacetime. Rather, they are local in the sense of representing different spacetimes but similar to a certain degree.

The map in question is the following:

$$\rho : T_{B,0} \longrightarrow H^1(K_0, T_{K_0}). \qquad (6.8)$$

More precisely, the map is said to be the Kodaira-Spencer map of the family of deformations (K, B, ϕ) at 0 – the value of the parameter $\lambda = 0$ corresponds to some central fiber of the family, namely, some arbitrarily chosen geometry $S \times K_0$ used as starting point of the deformations.

The map works in the following way: one feeds into the map some vector tangent to the parameter space B at 0. The output of the map is some first-order infinitesimal deformation K_{λ} of K_0 – $\lambda \in (-\epsilon, \epsilon)$. The deformation of the compact part is performed along the direction of the tangent vector fed. This map can be applied at each point λ of the parameter space B, also called base space of deformation.

In general, the base space B is a portion of the entire moduli space M of the theory, and $B_\epsilon \subset B$ is even smaller than B as it only contains first-order infinitesimal parameters around the point 0. That is, it only parameterizes first-order infinitesimal deformations K_λ of the central structure K_0.

Another way to put it is the following: the "infinitesimal" (or "local") structure of the moduli space M is what one may get by gluing together infinitesimally (or "locally") different spacetimes. Again, note well that, by construction, the topological structure of the moduli space cannot be the same as the topological structure of any of the individual spacetime composing the family. Here, one has a difference in kind: the topology of the moduli space is not of the ordinary kind, the one that can be found on ordinary spacetime.

On a more philosophical note, what the topological structure of the moduli space represents is the totality of possibility for spacetime geometry to be in a way or another.

However, we need to add on top of this topology some vehicle of information about the dynamics of some system studied against all these different geometrical backgrounds. So, like in some science fiction plot, the system will be allowed to travel from one spacetime to another, and the additional structure I'm going to define will keep track of the values taken by its physical properties as it travels from one spacetime to another.

Two more points. First, the moduli space described so far should be distinguished from the physical multiverse arising from the cosmological application of string theory. To be precise, our moduli space is a metaphysical structure (expressed in the formal language of deformation theory) depicting a space of possibilities. Once its construction will be complete in the next section, it will be used to point to the true metaphysical commitment of the theory about geometry. Indeed, it will show the different degree of background independence of the theory, depending on how one moves on that space. However, some of the parts of this moduli space depict genuinely nomological possibilities – included in the theory's laws. So, in this sense, the physical string theory multiverse gets reflected into the moduli space. By taking into consideration any smooth deformation of some geometry, it is not always the case that one can find a nomologically possible geometry – anyone of those dynamically produced by quantum strings. Here, the broader set of mathematical possibilities – which is replacing the traditional role played by the notion of logical possibilities – is the one obtainable from the theory of deformation applied to compact topological structures. And this set is inevitably bigger than the set of nomological possibilities dictated by quantum string theory.

Second, the existence of a Kodaira-Spencer map at $\lambda = 0$, i.e., (6.8), tells us that each tangent vector at $\lambda = 0$ to the moduli space identifies a possible deformation of the geometry K_0. If the Kodaira-Spencer map is surjective, then every first-order deformation of a geometry K_0 is represented by a tangent vector to the moduli space. So, under this condition of having a surjective map, a local linear approximation around the moduli point corresponding to K_0 picks an infinitesimal neighborhood of geometries around K_0 in K. As I said, this neighborhood depicts in the most general case a set of mathematical possibilities.

The condition under which the Kodaira-Spencer map is surjective is beyond the aim of this chapter.[18] However, if the map is surjective, the family of first-order deformations K is *complete*, namely, it contains all the possible first-order deformations of K_0 – mathematically and nomologically possible. So, in this context, the surjectivity of the Kodaira-Spencer map is equivalent to saying that there isn't any possibility that has been left aside.

6.3 Encoding dynamics: background independence

In this section, I complete the picture by defining on top of the topological structure some further structure carrying dynamical information. This type of information is about some set of physical properties of some physical string system, whose values are "taken" against different geometrical backgrounds and then "compared."

Let's consider some physical dynamics of a quantum system of strings, here denoted by Q, along a fixed background of propagation. Typically, such a system has many observables, which we can measure by computing their expectation values. Observables of a system are linear self-adjoint operators acting on the Hilbert space, the one representing the totality of the system's states. Let's denote with ψ a quantum state of Q, with H the Hilbert space of Q's quantum states, with O the vector space generated by the observables O_α of the system. One can think of the vector space O as a group of linear transformations O_α acting on H by mapping quantum states of the system onto different quantum states of the same system. Moreover, it is possible to define multi-linear maps over this space O that, depending on their ranks, map a certain number of observables onto their expectation values. Such multi-linear maps generate a vector space, here denoted by *Func*(O), defined in the following way:

$$Func(O) = \{h_n : O \longrightarrow \mathbb{C}; \forall n \in \mathbb{N}\},$$

such that

$$\forall \alpha, n = 1, h_1(O_\alpha) = < O_\alpha >,$$
$$\forall n > 1, h_n(O_{\alpha_1} O_{\alpha_2} \ldots O_{\alpha_n}) = < O_{\alpha_1} O_{\alpha_2} \ldots O_{\alpha_n} > .$$

Introducing observables of a system Q inside our picture requires the use of a specific type of algebraic objects, namely, fiber bundles over the moduli space M.[19] Before describing how these fiber bundles look like, let's briefly define what a fiber bundle is in general.

Definition 6.3.1. A fiber bundle with fiber F is a map $p: E \longrightarrow D$, where E is called the total space of the fiber bundle and D the base space of the fiber bundle. The main condition for the map to be a fiber bundle is that every point in the base space $d \in D$ has a neighborhood U such that $p^{-1}(U)$ is homeomorphic to $U \times F$ in a special way. More precisely, if q is the homeomorphism, i.e., q is defined by

$$q : p^{-1}(U) \longrightarrow U \times F,$$

then $proj_U \circ q = p_{|p^{-1}(U)}$, where the map $proj_U$ is the projection onto the U component of the Cartesian product. The homeomorphism q which "commutes with projection" is called local trivialization for the fiber bundle p. In other words, E, at least locally, looks trivial, i.e., like the product $D \times F$, except that the fibers $p^{-1}(d)$ for $d \in D$ may be a bit "twisted."[20]

Applying the above definition to our case, let's consider a string system Q along with the Hilbert space of its quantum states and with the space of its observables. The moduli space M of the theory describing the system is here the base space, and the fiber bundle over that base is $\overline{H}$, i.e., we have

$$\begin{array}{c} \overline{H} \\ \downarrow^{p} \\ M \end{array}$$

The fiber bundle $\overline{H}$ is defined as $\coprod_{\lambda \in M}(H_\lambda \times \mathbb{C})$. Each disjunct $H_\lambda \times \mathbb{C}$ is a fiber over a point λ of the base moduli space M. H_λ is the Hilbert space of Q's states whose dynamics are considered along the background parameterized by λ, and the vector space of complex numbers $\mathbb{C}$ represents all the numerical values assumed over λ by the transition functions $f_{\alpha_1 \alpha_2 \ldots \alpha_n}$ of the system. In other words,

$$f_{\alpha_1 \alpha_2 \ldots \alpha_n}(\lambda) = < O_{\alpha_1}, O_{\alpha_2}, \ldots, O_{\alpha_n} >,$$

where for any given λ, $f_{\alpha_1 \alpha_2 \ldots \alpha_n}(\lambda)$ assume the values of the functions $h_n(O_{\alpha_1}, \ldots, O_{\alpha_n})$ introduced above as I was considering the system over a fixed background.

So, $\overline{H}$ is the fiber bundle resulting from the disjunct union of all the possible Hilbert spaces H_λ of the system, each of them equipped with some extra structure. Such union is taken by varying over the family of backgrounds parameterized by λ.

Let's now define a section of the bundle. Let's consider a set of observables $\{O_{\alpha_1}, O_{\alpha_2}, \ldots, O_{\alpha_n}\}$ of the system Q. Recall that fixing a λ means fixing a background of propagation. By definition, we have that $\forall\ \lambda \in M$, $s(\lambda) \in (H_\lambda \times \mathbb{C})$. More precisely, the section of the fiber bundle maps a point λ of the base space M over a vector in the fiber in the following way:

$$s(\lambda) = (\psi_\lambda, f_{\alpha_1 \ldots \alpha_n}(\lambda)),$$

where ψ_λ is a quantum state of the system along the background λ and $f_{\alpha_1 \dots \alpha_n}(\lambda)$ is the correlation function value of the chosen set of observables, relative to that quantum state and that background. Let's also denote the vector $(\psi_\lambda, f_{\alpha_1 \dots \alpha_n}(\lambda))$ with v_λ.

Now, in order to say that this topological and dynamical structure on the theory's moduli space points to background independence, one would need to show that, given a set of observables like above and an arbitrarily picked member λ_0 of the family, the value $f_{\alpha_1 \dots \alpha_n}(\lambda_0) = < O_{\alpha_1} O_{\alpha_2}, \dots, O_{\alpha_n} >$ remains constant over all backgrounds of the family, i.e., $\forall\ \lambda \in M$,

$$f_{\alpha_1 \dots \alpha_n}(\lambda_0) = f_{\alpha_1 \dots \alpha_n}(\lambda).$$

Now, to study in which cases the transition functions' values remain constants, one needs a way to compare vectors lying in different fibers of the vector bundle over two different points λ_1 and λ_2. In other words, one needs to pick a *flat connection* and some path over M from λ_1 to λ_2, along which a vector can be dragged in the fiber over λ_1 all the way to the fiber over λ_2. Once there, the dynamical information the vector carries can be compared with that contained in the fiber over λ_2. In other words, we need to make a "parallel transport" of information.[21] Let's dig into this issue.[22]

Let $[0, T] \longrightarrow M$, $t \in [0, T]$, be a smooth path $\lambda(t)$ over the moduli space M from λ_1 to λ_2, i.e., $\lambda_1 = \lambda(0)$ and $\lambda_2 = \lambda(T)$. Let's denote with $v_{\lambda(t)}$ the vector in the fiber of $\overline{H}$ over $\lambda(t)$. We need to say that $v_{\lambda(t)}$ is "parallel transported" along $\lambda(t)$. That can be done by finding an equation involving the covariant derivative of $v_{\lambda(t)}$ in the direction $\lambda(t)$ is going, i.e., $\lambda'(t)$. Since we need to use a flat connection, our equation comes from the vanishing of the covariant derivative,[23]

$$D_{\lambda'(t)} v_{\lambda(t)} = 0. \tag{6.9}$$

So, we can say that $v_{\lambda(t)}$ is parallel transported along $\lambda(t)$ if and only if the above covariant derivative vanishes $\forall\ t \in [0, T]$.

The differential equation above can always be solved, and if one gives an initial condition over λ_1, the solution is unique. This solution is the vector with which we wind up in the fiber over λ_2. At this point, we should recall two things.

First, $\forall \lambda$, v_λ in our case is $(\psi_\lambda, f_{\alpha_1 \dots \alpha_n}(\lambda)) \in H_\lambda \times \mathbb{C}$ and $\{O_{\alpha_1}, O_{\alpha_1}, \dots, O_{\alpha_1}\}$ is a set of observables of the system.

Second, fixing an initial condition $v_\lambda(0) = v_{\lambda_1} = (\psi_{\lambda_1}, f_{\alpha_1 \dots \alpha_n}(\lambda_1))$, we would like that the solution v_{λ_2} is a quantum state of the system over the background λ_2 preserving all the expectation values assumed over λ_1 by the set of observables mentioned above. This requirement, if satisfied for any arbitrarily fixed initial condition λ and for any observable of the system, would support the strongest form of background independence.[24]

Now, if on one side the existence of a solution is guaranteed, on the other side, it is not true in general that the solution respects the above requirement. We can always have a flat connection over a general fiber bundle, the one representing the canonical way of dragging along a path some vector belonging to the fiber of some point. Eventually, the original vector will end up in some fiber on top of some other point, and there it will be compared with some other vector already sitting in that fiber. But there isn't any general rule dictating that the expectation values will be surely be preserved.

However, it turns out that the preservation is respected as the string system is travelling through geometrically inequivalent spacetimes, in so far as they are topologically equivalent. This is virtually what we get out of T-duality. But one needs to be careful here and to understand to what extent the topological equivalence plays a role. The case analyzed in the bosonic T-duality was a simplified instantiation of topological equivalence between two types of compact extra dimensions. That was a case in which the physical properties of a string system Q are insensitive to a change of the background radius from r to $\frac{1}{r}$. In particular, the observables $\{O_{\alpha_1}, O_{\alpha_2}, \ldots, O_{\alpha_n}\}$ sensitive to the total energy of the string configuration do not detect the difference, and so we have the following preservation:

$$f_{\alpha_1 \ldots \alpha_n}(\lambda_r) = f_{\alpha_1 \ldots \alpha_n}(\lambda_{\frac{1}{r}}) = < O_{\alpha_1} O_{\alpha_2} \ldots O_{\alpha_n} > .$$

So, one arrives at the formulation of a first claim of string theory background independence. Let's call it Thesis 1: any family of geometrically different spacetimes, if the topological invariants are preserved, can be taken as the data for constructing a string theory without any prejudice to choice of a particular member.

In the simplified case of T-duality presented in Chapter 4, the topological invariants are preserved as one shrinks the radius r of the compact extra dimension to the radius $\frac{1}{r}$. Note well that, if one shrinks the radius r to an arbitrary smaller value s, different from $\frac{1}{r}$, T-duality does not apply. What I mean is that the T-duality orbits of physical invariance over the moduli space are discrete; each orbit contains two elements – a theory of radius r and its T-dual of inverse radius. One cannot rely on T-duality to claim that along the path on the moduli space from $\frac{1}{r}$ to r, the dynamical behavior of the string system remains unaffected. Now, to claim that the string dynamics are preserved via an arbitrary shrink (or dilatation) from r to s, one might propose a different argumentative line that does not involve duality but involves conformal invariance. As we will also see in the string mirror symmetry case, one might informally say that the perspective of the string theory's moduli space is internal to the theory – whether this abstract space is construed as discussed in this chapter or by sticking to more conventional physical applications. Such internal perspective is identifiable to that of a two-dimensional field theory which is conformally invariant and whose coupling parameters encode the topological features of the compact part of the background. Now, any two-dimensional CFT is insensitive to arbitrary shrinking or dilating. So, in this sense, the preservation of the expectation values above for any arbitrarily picked pair of radii would be a byproduct of the insensitivity of the moduli space's perspective.

6.4 Duality without preservation of topological invariants

If string mirror symmetries are involved, one might formulate a stronger claim of background independence. Let's call it Thesis 2: any family of geometrically different spacetimes, all sharing vanishing Ricci curvature, and differing in a certain way for some topological invariants, can be taken as the data for constructing a string theory without any prejudice to choice of a particular member. Let's consider an instantiation of this claim.

The following diagram summarizes the formal construction made so far:

$$\begin{array}{lllllll} \overline{H} & & & & & & \\ \downarrow^{p} & & & & & & \\ M & \xleftarrow{p_2} & S\times M & \xleftarrow{id\times\phi} & S\times Def(K_0) & \xleftarrow{i} & S\times K_0 \end{array}$$

$$p^{-1}(\lambda) = (\psi_{\lambda}, f_{\alpha_1\ldots\alpha_n}(\lambda)) \in H_{\lambda}\times\mathbb{C}.$$

Let's reformulate the Vafa's definition of duality in light of the formal language introduced in this chapter: two distinct string theories are dual just in case (1) they have isomorphic moduli spaces and (2) there is an isomorphism between observables compatible with all the correlation functions. Let's then consider two dual string theories, namely, $T[M, S\times Def(K_0), O_\alpha]$, and $T'[M', S'\times Def(K_0'), O_\beta]$.

As I said, in string theory, the topological invariants of the compact extra dimensions contribute to produce the low-energy physics in the non-compact spacetime. What allows quantum string to make the best possible contact with observed particle physics is a specific set of compact shapes, that is, the Calabi-Yau manifolds having Euler characteristics equal to ±6. As a matter of fact, a string theory with Calabi-Yau extra dimensions having Euler characteristics equal to +6 finds its dual in another string theory with Calabi-Yau extra dimensions having Euler characteristics equal to −6. The duality in this case is called mirror symmetry.[25] Because of the different Euler characteristic, the compact parts of the backgrounds are topologically inequivalent.

Now, it is possible to express the mirror symmetry relation between the two theories via the diagram introduced above. One might start from the relation between the spacetimes posited as background of propagation in each mirror symmetric theory. The two spacetimes are in correspondence through the mirror mapping ψ_0 that changes the sign of the Euler number,

$$S \times K_0 \longrightarrow^{\psi_0} S' \times K_0'.$$

Now, each background is smoothly deformed:

$$\begin{array}{ccc} S \times Def_\lambda(K_0) & \longrightarrow^{\psi} & S' \times Def_{\lambda'}(K_0') \\ \uparrow^{i} & & \uparrow^{i} \\ S \times K_0 & \longrightarrow^{\psi_0} & S' \times K_0'. \end{array}$$

Loosely speaking, the smoothness of the deformations of the central fiber K_0 preserves its Euler number inside the family and allows the map ψ_0 to be lifted to the map ψ which applies to every fiber of the family and maintains the property of being a mirror mapping.

Now, the families of deformations produced on both sides end up with being depicted by the local structures of the relative moduli spaces:

$$\begin{array}{ccc} \overline{H} & & \overline{H'} \\ \downarrow^{p} & & \downarrow^{q} \\ M & & M' \\ \uparrow^{\phi} & & \uparrow^{\varphi} \\ S \times Def_\lambda(K_0) & \longrightarrow^{\psi} & S' \times Def_{\lambda'}(K_0') \\ \uparrow^{i} & & \uparrow^{i} \\ S \times K_0 & \longrightarrow^{\psi_0} & S' \times K_0'. \end{array}$$

By following a similar argumentative line to that used in the previous section, one might take the following argumentative step. Although the two families of deformations $S \times Def_\lambda(K_0)$ and $S' \times Def_{\lambda'}(K_0')$ are topologically inequivalent because of the mirror maps, from the point of view of their moduli spaces, that inequivalence vanishes. This fact might be explained as before, by saying that the moduli space's perspective is internal to the string theory related to that space. For this reason, this internal perspective can be identified to that of a two-dimensional CFT equipped with coupling parameters encoding the backgrounds' topological properties. CFT is insensitive to the mirror mapping. So, this is why in the diagram below we can lift the map ψ to the isomorphisms $\psi^\star$ and η as depicted:

$$\begin{array}{ccc} \overline{H} & \longrightarrow^{\eta} & \overline{H'} \\ \downarrow^{p} & & \downarrow^{q} \\ M & \longrightarrow^{\psi^\star} & M' \\ \uparrow^{\phi} & & \uparrow^{\varphi} \\ S \times Def_\lambda(K_0) & \longrightarrow^{\psi} & S' \times Def_{\lambda'}(K_0') \\ \uparrow^{i} & & \uparrow^{i} \\ S \times K_0 & \longrightarrow^{\psi_0} & S' \times K_0'. \end{array}$$

The map η on top of the diagram is the lift of the isomorphism $\psi^\star$ between the two moduli spaces. The former is the isomorphism between the structures of fiber bundle on each moduli space, each structure given by the theories' observables. Precisely, η is the isomorphism between the two fiber bundles $\overline{H}$ and $\overline{H'}$ in virtue of which $\forall(\psi_\lambda, f_{\alpha_1 \ldots \alpha_n}(\lambda)) \in \overline{H}$,

$$\exists!\eta(\psi_\lambda, f_{\alpha_1 \ldots \alpha_n}(\lambda)) = (\phi_{\psi^\star(\lambda)}, f_{\beta_1 \ldots \beta_n}(\psi^\star(\lambda)) \in \overline{H'},$$

such that the following compatibility condition is respected: $\forall\ \lambda \in M$ and $\forall\ n \in \mathbb{N}$,

$$f_{\alpha_1 \ldots \alpha_n}(\lambda) = f_{\beta_1 \ldots \beta_n}(\psi^\star(\lambda)).$$

The fact that the fiber bundle map η is an isomorphism between the two fiber bundles, which satisfies those compatibility conditions, shows how the mirror symmetry switching the Euler characteristic (at the bottom of the diagram) preserves the physical and dynamical content of the two string theories T and T'.

6.5 Are deformation of laws more fundamental than laws?

So far, I explored different string dualities and how they get differently reflected into the local structure of the theory moduli space. Informally speaking, each duality corresponds to some set of equivalence orbits on top of the moduli space, where the equivalence relation here can be described as "having the same expectation values for any observable of any system." Then, one might have a better sense at this point of what it means to say that string theory shows different degrees of background independence. Without pushing further any claim about background independence, I would like to shift the attention on something deeply correlated. If the argumentative lines presented so far are convincing, what can be said about the fundamental structure of the world according to quantum string theory?

One might start with saying that a potential candidate should be a relational structure, something weaker than a geometry. At the beginning of this chapter, I said that the "act" of deforming a geometry, or more generally a topology, might be more fundamental than the latter. The rationale behind this claim resides in the fact that deformations of spacetime geometries constitute the topology of the moduli space. An infinitesimal neighborhood in this topology arises from infinitesimally deforming some geometry. The less continuous the deformations are, the wider the corresponding neighborhood on the moduli space is. And so the several degrees of similarity one can maintain when deforming a geometrical structure produce a topological notion of locality on the moduli space.[26] But why should this topology be more fundamental than any ordinary one? One might say that all those cases of string physics' insensitivity to the lack of preservation of topological invariants might answer the question straightforwardly. But as we saw, string dualities not always come with such insensitivity. So, a general answer is actually not straightforward.

As I said, a topology here is characterized by the formal property of being a relational structure, something that formally represents interactive dynamics of string systems. Based on that, I also said that a claim about topology is a claim about dynamical laws. Now, if one can extract a topology (that of the string theory moduli space) from the deformations of some topological structure (any physical spacetime), then it is like one can extract some sort of metalaw from the deformations of some law.

Notes

1 In Section 4.3.2, I explained what "weaker" means in this context and how topologies can be non-geometrical.

2 Note well that here the sole function of coordinate systems is to bring differentiability into play. Coordinate systems can be used to solely represent a differentiable structure, namely, something purely topological, hence weaker than a metrical (or geometrical) structure. This point is clearly explained by Mühlhölzer in his "Scientific Explanation and Equivalent Descriptions": differentiable structures must be carefully distinguished from metrical (geometrical) structures. Any differentiable structure is represented by a class of coordinate systems. And the function of coordinate systems does not necessarily bring metrics into play. The geometrical structure on the other hand is represented by a tensor field on the differentiable structure as a whole and not by distinguished coordinate systems. Then, coordinate systems can have a double function, namely, to represent the differentiable and the metrical structure – Felix Mühlhölzer in "Logic, Language, and the Structure of Scientific Theories: Proceedings of the Carnap-Reichenbach, Centennial", edited by Wesley C. Salmon, Gereon Wolters, University of Pittsburgh Press, 1994, pages 126, 127.

3 String mirror symmetries are briefly described in this chapter. Note that here I am keeping out of the story the holographic duality. The latter presents an even bigger obstacle since it delivers a much more radical scenario of topological discontinuity between the two dual descriptions, not to mention that one of them does not even contain strings.

4 See also Rickles (2010, page 60) and Vafa (1998, page 540).

5 The rich representative role of these spaces make them flexibly applicable to a heterogeneity of contexts in advanced physics. But often they also appear in many familiar mathematical contexts. Imagine the ordinary two-dimensional set of Cartesian coordinates (x,y), where x and y are two variable real numbers. Commonly denoted by $\mathbb{R}^2$, this numerical set is frequently used to *represent* a set of geometrical objects, namely, points in a plane. Then, one can legitimately say that $\mathbb{R}^2$ is a moduli space representing points in a plane.

6 A possible scenario arising from discontinuous deformations changing geometry, topology, and the number of dimensions is that characterizing the AdS/CFT correspondence seen in Chapter 5. The AdS/CFT duality corresponds to some "path" of physical invariance over the moduli space. The path would connect areas of this space much more far apart than those connected by paths corresponding to T-duality.

7 Note well though that this separation between the topological structure and the dynamical one is a huge oversimplification. Indeed, it is often the case that those structures of the moduli space in charge of reflecting dynamics (for example, coupling constants) depend on those in charge of encoding geometry.
8 A holomorphic function (or map) is a complex-valued function of one or more complex variables that is complex differentiable in a neighborhood of every point in its domain. The topological and functional structures we mentioned above are all defined in relation to the domain of complex numbers. Any topological structure in the family (in virtue of its local charts) is everywhere locally regarded as being a complex numerical space. This makes it possible to use functional calculus on it.
9 What follows about deformation is based on the content of "Complex manifolds and deformation of complex structures" by Kunihiko Kodaira, see Kodaira (1986, pages 182–208). Also see pages 265–266 of "Algebraic geometry" by Robin Hartshorne (Hartshorne, 1977), along with "Lectures on deformations of complex manifolds" by Marco Manetti (Manetti, 2004).
10 I am following here Kodaira's approach, see Kodaira (1986, pages 182–192).
11 See Kodaira (1986, pages 182–192). In particular, see definition 4.1, page 184.
12 $n = dim_{\mathbb{C}}(K_\lambda)$.
13 This endnote provides some broader context to read what we just said about local charts and gluing functions. A complex manifold K_0 is a manifold equipped with a complex structure which is defined by a system of local complex coordinates $(z^1,\ldots,z^n)$ around each of its points. More precisely, for all $p \in M$, we can find a coordinate open neighborhood V_j in which p is determined uniquely by its local coordinates $z_j = (z_j^1,\ldots,z_j^n)$. The manifold should be thought as entirely covered by such collection of open neighborhoods $\{V_1,\ldots, V_j,\ldots\}$. On each V_j and $\forall\, p \in V_j$, a homeomorphism $z_j : p \longrightarrow z_j = (z_j^1,\ldots,z_j^n)$ is defined. Then, for each pair j,k with $V_j \bigcap V_k \neq \emptyset$, the map $f_{jk} : z_k(p) \longrightarrow z_j(p)$, $p \in V_j \bigcap V_k$, is a biholomorphic coordinate transformation. The open charts V_j glue together through these biholomorphic transformations. This fact guarantees that they glue consistently with the requirement that p is uniquely determined by its local coordinates. Ibid., page 29.
14 In fact, such metrics always exist in abundance on any complex manifold.
15 For an exhaustive and introductory presentation of the notions of sheaves, cohomology groups, and 1-cocycle, I refer the reader to Kodaira (1986, pages 109–133).
16 In differential geometry, a 1-form on a differentiable manifold is a smooth section of the cotangent bundle. Equivalently, a 1-form on a manifold M is a smooth mapping of the total space of the tangent bundle of M to M, whose restriction to each fiber is a linear functional on the tangent space.
17 This is a useful technical note if one wants to dig a bit into the formal aspects of the story: indeed the sheaf of functions over B_ϵ is equal to the quotient $\frac{O_{B,0}}{m^2_{B,0}}$, where $m^2_{B,0}$ consists of functions vanishing at 0.
18 For detail on that, a simple exposition is in "Lectures on deformations of complex manifolds" by Marco Manetti (see Manetti, 2004, page 12)
19 See also Vafa (1998, page 540, endnote 2).
20 For more detail see http://mathworld.wolfram.com/FiberBundle.html.
21 What follows about connection heavily relies on "Gauge fields, knots and gravity," by John Baez and Javier Muniain (2008, pages 223–242). For an introduction to the preliminary notions on which my presentation relies, I refer the reader to this book.

22 See also Baez and Muniain(2008, pages 233–235).

23 For a definition of covariant derivative, see Baez and Muniain (2008, pages 223–229). Here, such derivative is defined for a section s of the fiber bundle. However, from page 233 to page 234, Baez using the idea of a section s as vector-space-valued function redefined by analogy the covariant derivative of a vector in a fiber along the path over the manifold, i.e., the derivative I introduced above.

24 Why is that? Because by construction, any parameter λ picks some geometry.

25 For a surface of genus g, the Euler characteristic is defined like $\chi(g) = 2 - 2g$.

26 Elsewhere I attempt to revise the Lewisian similarity order in light of the topological language of deformations applied to string theory – "Modal realism after String Theory," in "Beyond Spacetime: The Philosophical Foundations of Quantum Gravity," Cambridge University Press, forthcoming.

7 Spacetime emergence via non-commutative quantum string theory

In this chapter, spacetime emergence in quantum string theory does not involve string dualities. Rather, the angle from which I make my inquiry is given by non-commutative geometry. The main idea is that spacetime might arise as an emergent structure from some underlying non-commutative structure that quantum string theory might account for. A potential feature of quantized space and quantized time might be that of failing to commute. In the first section, I will study the implications of space-time non-commutativity for the causal structure of quantum string theory. Since a causality talk is still possible in this non-commutative setting, one might find a perspective from which the theory's background independence can be argued along an argumentative line different from the main one followed so far. Then, in the second section, I shall analyze the nature of the conditions under which space-time non-commutativity arises in quantum string theory. Does this behavior appear just in perturbative regimes or is it instead an intrinsic property of the theory? Two different approaches on this issue will be presented.

7.1 Is causality still in the theory?

The main question underlying this section is: does space-time non-commutativity affect the causal structure of quantum string theory?[1] I start considering the issue of causality in the perspective of non-commutative field theories. Some of these field theories arise as low energy limits of non-commutative string theory, whereas some others cannot arise in this way. Both cases provide us with crucial information about whether causality is preserved in non-commutative string theory.

Non-commutative field theories are field theories over a spacetime equipped with a non-commutative geometry. The ordinary notion of space, the one in which spatial coordinates commute, is usually described by the Euclidean metric. Sticking to this conventional choice, I describe in basic terms what a non-commutative geometry is. For simplicity reasons,

I provisionally consider a two-dimensional space. A two-dimensional Euclidean space is a plane equipped with Euclidean metric, namely, a metric induced by ordinary "·" multiplication.[2] In this space, the computation of the area A of a rectangle of sides x^1 and x^2 does not depend on the order of the factors in multiplication, namely, $x^1 \cdot x^2 = x^2 \cdot x^1 = A$.

An algebraic generalization of the ordinary multiplication is the ⋆-product defined by the following:

$$x^1 \star x^2 - x^2 \star x^1 = \theta^{12},$$

where θ^{12} is a non-zero antisymmetric parameter. So, ⋆ fails to be commutative. In this case, the order of multiplication matters since $x^1 \star x^2 = x^2 \star x^1 + \theta^{12}$. Then, which kind of metrical structure, if any, will be induced over the space by the ⋆-product? In a space like that, computing areas would become an impossible task. Notions like length, area, and volume are devoid of meaning as any ordinary metric disappears. Indeed, the ⋆-product introduces a more abstract notion of space, namely, that of an *algebraic space*. This abstract notion, entirely based on some purely algebraic relational structure, is a generalization of the ordinary one. Non-commutative geometry is the mathematics that allow us to work with non-geometric spaces.

Mutatis mutandis, a non-commutative spacetime of dimension D works exactly like the two-dimensional case. In a D-dimensional spacetime, non-commutativity can be either purely spatial or space-time non-commutativity. Some interesting comparative studies concerning spacetime non-commutativity in field theory and string theory are presented by two papers of Seiberg, Susskind, and Toumbas along with a paper of Gomis and Mehen.[3]

In these works, two classes of non-commutative field theories are formally derived from a commutative field theory by replacing the ordinary field product with the ⋆-product defined above. The antisymmetric parameter $\theta^{\mu\nu}$ is more precisely an antisymmetric matrix, and it is defined by the non-commutation relation among coordinates in the following way[4]:

$$[x^\mu, x^\nu] = x^\mu \star x^\nu - x^\nu \star x^\mu = i\theta^{\mu\nu}, \tag{7.1}$$

where $\mu, \nu = 0, \ldots, D-1$ and with D equal to the spacetime dimension.[5] The first class of non-commutative field theories is characterized by just space-space non-commutativity. In this case, the antisymmetric parameter is such that $\theta^{0i} = 0$. The second class is instead characterized by space-time non-commutativity, i.e., $\theta^{0i} \neq 0$. This first group of findings is compared by the authors with a second group of results concerning string dynamics along non-commutative backgrounds.

Before analyzing these results, I want to make few remarks on the mentioned papers. My view of the conceptual relation between ordinary space-time commutative theories and space-time non-commutative theories does not seem to match theirs. Perhaps they would not agree with my conceptual framework in which emergence of an effective commutative spacetime theory is seen as result of specialization of a more general, underlying, non-commutative one.

Indeed, in their approaches, things seem to go in the other way around: a space-time non-commutative theory arises from a commutative one as a particular case obtainable by perturbating the latter's background with an electric field. However, the authors do not seem to endorse the idea that ordinary spacetime is a fundamental structure. Indeed, it is unambiguously considered to be an effective entity emerging from some deeper underlying structure. Yet in those works, space-time non-commutativity is not explicitly considered to be a feature of the presumed underlying structure. Maybe I should not be reading too much in this lack of explicit assignment of metaphysical status to space-time non-commutative structures. It might well be possible that their methodological choice to start with a commutative spacetime for then perturbating it does not reflect any metaphysical endorsement about the status of non-commutativity. If instead the methodological choice reflects some metaphysical thesis, the latter might have to do with the conceptual status assigned to a principle that plays a central role in this debate, i.e., the space-time uncertainty principle – this will be the topic of Section 7.2.

Now, let's unravel the content of these findings. First, I shall describe briefly the case of space-space non-commutative field theory and what kind of information that delivers about space-space non-commutative string theory. Second, following the same pattern, I will analyze more extensively the case of space-time non-commutativity.

In the case of space-space non-commutativity, the field theory is not local in space but still local in time. This kind of non-locality destroys the Lorentz invariance of the theory. In fact, the Lorentz transformations acting on this type of non-commutative space end up with being applied also to the new background field represented by $\theta^{\mu\nu}$, since this parameter has Lorentz indices.[6]

As Carroll et al. (2001) claim, there are two different types of Lorentz transformation.[7] The first one is the rotation of the "observer inertial frame," which does not change the physics because the field operators in the Lagrangian and the $\theta^{\mu\nu}$ are invariant under them. The second one is the "rotation of a particle" inside a "fixed observer frame."[8] The field $\theta^{\mu\nu}$ is not invariant under their action, and this fact produces a different physics.

So, more precisely, any space-space non-commutative field theory violates a particular kind of Lorentz symmetry. However, breaking this symmetry does not entail a lack of unitarity of the theory, and so it does not raise the specter of indeterminism.[9]

The presence of $\theta^{\mu\nu}$ does not change the rules and the usual framework of quantum mechanics. In fact, locality in time of the action allows the construction of a Hamiltonian that yields a unitary time evolution.[10] The unitary structure of these field theories provides us with crucial information about string theory because they arise as low energy limits of string theory in the presence of a background magnetic field.[11]

The low energy limit is formally implemented by the limit $\alpha' \to 0$ – where α' is the string parameter – because the only dimensionless parameter in the theory is $(\alpha' E^2)$.[12] In this low energy limit, the dynamics are described by the non-commutative field theory of the massless open strings (the massless modes are all we can observe in that regime because the massive ones are too heavy to be seen). So, the fact that this limit can be performed tells us that space-space non-commutativity preserves the deterministic structure of string theory as well. In other words, the effective field theory comes out from a "consistent truncation of the full unitary string theory."[13]

Things are different for space-time non-commutative field theories ($\theta^{0i} \neq 0$). The non-commutative behavior of time in these field theories causes two important anomalies. The first one is the presence of acausal behavior at the first perturbation level (tree level) – see Seiberg, Susskind, and Toumbas (2000b). The second one is the arising of a non-unitary time evolution of the fields, i.e., a failure of determinism, at the second perturbation level (one-loop level) – see Gomis and Mehen (2000) and Seiberg, Susskind, and Toumbas (2000b).

The physical process I heuristically present here is taken from Seiberg, Susskind, and Toumbas (2000a, 2000b). This is the scattering process of localized scalar fields or, in other words, the scattering of the wave packets of high-energy particles after a collision.[14] Here, I will use the same reduced number of dimensions (one spatial and one temporal) to make the description easier. Also, writing just θ is a shorthand for $\theta^{\mu\nu}$.

Before the collision, we have two incoming particles having momenta respectively k_1 and k_2; after the collision, we have two outgoing scattering particles having momenta respectively p_1 and p_2. Using perturbation theory, it is possible to calculate the S-matrix relative to this interaction.[15] Now, an important fact to emphasize about the process described here is that the collision of the two incoming particles is the cause of the scattering, since the particles only interact at that time; before and after the collision, the system is free. This remark will be relevant for what follows about acausal behavior that involves the first perturbation level.

In a commutative theory, the S-matrix entries contain an invariant amplitude Ω, which at the first level of perturbation is equal to "$-ig$," where g is a coupling constant appearing in the interaction term of the Lagrangian.[16]

By switching to the non-commutative case, the ordinary product among fields is replaced by the $\star$-product defined above which is defined in function of the non-commutativity parameter θ. This new product brings several changes in the interaction term of the system's Lagrangian. Indeed, the interaction term will appear like[17]

$$g\phi \star \phi \star \phi \star \phi. \tag{7.2}$$

Therefore, this $\star$-product introduces an additional new phase in the interaction, called Moyal phase, which depends on the energies p^2 of the outgoing particles and on the θ parameter. This can be also seen from the entries of the new S-matrix which contains the new amplitude $\Omega = g(f(p^2\theta))$, where $f(p^2\theta)$ is a periodic function. The outgoing wave function, hence, will contain this new phase[18]:

$$\Phi_{out}(x) \sim \int dp(f(p^2\theta))\phi_{in}, \tag{7.3}$$

where ϕ_{in} is the incoming wave packet.

Integrating over the momenta dp, we will get an outgoing wave packet – after the collision – that splits in three parts, but only two are relevant to us. The first one appears time-delayed with respect to the collision of the incoming packet. It is called the *retarded* wave packet. The second outgoing term is instead an *"advanced"* wave packet since it appears before the collision. The former represents a physical process compatible with causality, whereas the latter represents an acausal physical process, also considered by the authors as reason to reject a space-time non-commutative field theory as pathological.

But is this acausal behavior an instance of backward causation? If that is the case, it should not be seen as a violation of the physical laws governing the process. Fundamental physical laws are time-symmetric after all. The fact that the cause always precedes the effect is not a lawful fact. It is something that should be explained independently from the laws.

An example from classical electrodynamics can help to explain more clearly this point. Learning to compute a solution in this context reveals to us that a wave equation is formally associated with both the retarded G^+ and advanced G^- Green functions. However, any actual field configuration (solution to the wave equation) can be constructed selecting just one of the two Green functions. Physicists usually choose the retarded Green function. This function is usually thought as the "causal" one. However, selecting the advanced Green function would be an equivalent choice.

There is nothing special about the retarded Green function that makes it consistent with causality. That choice just depends on the particular Cauchy problem we are dealing with. The notion of cause in classical electrodynamics does not include the requirement that the cause must precede the effect. A cause is a physical event – usually an interaction – which is formally described by a Green function that can be chosen either advanced or retarded.

Therefore, *mutatis mutandis*, we could just apply the same pattern of interpretation to the appearance of our advanced wave packet, hence concluding that it does not violate the physical laws governing the scattering process. But in this case, that appearance cannot be read as an instance of backward causation in the field theory case. Why is that?

Let's think for a moment to how an ordinary scattering process – i.e., a process following the ordinary temporal sequence – would appear. In this case, we would see an incoming wave packet corresponding to two increasingly close incoming particles moving from $t = -\infty$ to $t = 0$, the collision time. Then, for $t > 0$, we would see the effect of the collision, i.e., an outgoing wave packet corresponding to two increasingly distant outgoing particles moving toward $t = \infty$. For simplicity reasons, imagine a watch traveling along with the center of mass of the particles system with its hands moving clockwise.

Now, if such system takes part in a backward causation process, what we would see is the watch's hands starting to move counterclockwise, since the time sense of the system would appear reversed. So, we would see an advanced wave packet – the advanced effect – now corresponding to two increasingly near incoming particles, traveling backward from $+\infty$ toward $t = 0$. Then, for $t < 0$, we see now two increasing distant outgoing particles, traveling backward toward $-\infty$. In this case, we can say that the effect appears as advanced with respect to the cause – the collision – and the entire process appears reversed.

But this is not what happens in our scattering process. As Seiberg, Susskind, and Toumbas[19] crucially point out, the advanced part of the outgoing wave packet (the advanced effect) propagates toward $+\infty$. So, the "advanced" effect here consists in two outgoing increasingly distant particles – both moving toward $+\infty$ – appearing before the collision. That means that a watch moving along with them would keep rotating its hands clockwise. But if the time sense of the system is preserved, we cannot apply the notion of backward causation.

This violation of causality seems to have a different and problematic status. The authors try to explain this acausal behavior in terms of a similarity with an apparent advanced process in the dynamics of a rigid rod thrown against a wall from an initial distance.[20] The rigid rod has a center of mass moving along with it. One of its ends will eventually hit the wall before the center of mass does the same. So, the center of mass will appear to bounce back before it hits the wall. As they claim, the rigidity implies that the effect has a space-like cause.

But then, this process is impossible because it violates relativity. That would label the theory as pathological.[21]

In order to introduce the further problem arising at the one-loop level in the scattering case, I refer back to what I just said about Green functions and actual field configurations in electrodynamics. The reader should keep in mind that in what follows about one-loop level in the scattering case, the notions of advanced and retarded will be used in the sense of classical electrodynamics.

Although computing a solution involves an arbitrary choice of one of the two Green functions, the actual configuration of an electric field at any fixed time should not contain both functions (advanced and retarded with respect to the fixed time in which we look at the actual configuration). Equivalently, the actual configuration of the electric field at a present time should not have a value depending both on the sources in space at a past time and on the same sources in space at some future time.

That would be a simultaneous dependence of the present field configuration on two different initial conditions. But two different initial conditions cannot generate two solutions intersecting each other somewhere. That would violate determinism. Here, I am identifying the notion of deterministic evolution of a physical system with the mathematical property that an operator describing that evolution must have, i.e., being a one-to-one map among states.

Referring to the scattering field case, we have first to translate what we just said about violation of determinism in the language of quantum phenomena. In quantum physics, the notion of deterministic evolution of a system is often given via the definition of unitarity. Unitarity is a restriction on the allowed time evolutions that a quantum system can possibly have. Time evolution has to be mathematically described by a unitary operator, as a result of which probability is conserved. Unitary operators are automorphisms of Hilbert spaces, i.e., they preserve the linear space structure and the inner product of the space on which they act. In particular, since they are automorphisms, they have the property of being one-to-one maps between states.

So, the mathematical property of describing deterministic evolutions of the system is built inside the mathematical definition of being a unitary operator. So, in this sense, violation of determinism via violation of the mathematical requirement of being a one-to-one operator is a violation of unitarity. An example of that would be the simultaneous dependence of the present state's value of a system on both future and past states' values.

What we just said should clarify in which sense Gomes and Mehen speak about violation of unitarity at the second (one-loop) level of perturbation. In their "Space-Time Noncommutative Field Theories And Unitarity" (see Gomis and Mehen, 2000), they analyze the same field scattering case studied by Seiberg, Susskind, and Toumbas (2000b). They show that, since time does not commute ($\theta^{0i} \neq 0$), the Lagrangian contains non-local time derivatives, which makes the theory non-local in time.[22] The non-locality of the action produces at the one-loop level an actual configuration of the field in which its value at a present time depends simultaneously on both past and future times (see also Seiberg, Susskind, and Toumbas, 2000b, page 1).

Therefore, at the one-loop level, the advanced and the retarded wave packets appear together in the same outgoing wave function. Formally speaking, this simultaneous presence seems to be due to the appearance of the extra phase containing the antisymmetric parameter θ mentioned above (the Moyal phase). Therefore, space-time non-commutativity in field theory causes a failure of unitarity in the sense specified above.

Now let's consider what happens if we have space-time non-commutativity in a case of scattering strings. Why is this result about non-commutative space-time field theories relevant to the string case?

The interesting fact is the following: although space-time non-commutativity, that is ($\theta^{0i} \neq 0$), can be obtained in string theory, there is no way to obtain non-commutative space-time field theories as low energy limits of string theory.[23]

Why is that? Because space-time non-commutativity in string theory has implications for causality which are different from those found in field theory. I shall briefly describe the scattering process in non-commutative open string theory presented by Seiberg, Susskind, and Toumbas (2000a, 2000b, pages 9–14).

Before considering the scattering case studied by the three authors, let's mention some mathematical aspects involved in the presence of an electric field along a background of string's propagation. Let G be the metric of the open strings, $\theta^{0i} = \theta$, let E be the background electric field, and let E_c be the critical upper bound value of the electric field. The critical upper bound is the value of the electric field beyond which the string perturbative regime breaks down.[24] All these parameters are related to each other by the following relation[25]:

$$\alpha' G^{-1} = \frac{1}{2\pi}\frac{E}{E_c}\theta. \tag{7.4}$$

Intuitively speaking, we can see in this relation that $\alpha' \sim \theta$, for finite values of G. A low-energy limit of an open string theory is a limit in which $\alpha' \rightarrow 0$. This limit then implies also the vanishing of the non-commutative parameter. Therefore, what would seem to arise in this case is an ordinary commutative field theory. From the formula above, we can see that since $\alpha' \sim \theta$ and $E \sim \frac{1}{\theta}$, the energy scale that allows space-time non-commutativity to appear is the scale at which $\theta >> 0$. But then in this regime, the massive open string states cannot be neglected.

Now, this basic mathematical explanation of why we cannot perform the above-mentioned low energy limit should be pulled alongside of a physical explanation. To this aim, let's move to the authors' presentation of the string scattering case.

We just saw that in non-commutative field theory, the $\star$-product among fields changes the interaction term of the Lagrangian in such a way that a new phase depending on θ appears.

This new phase is responsible for a field scattering process that violates causality.

In the case of open string scattering, a new feature due to the oscillation of the strings changes the non-commutative parameter θ in the following way[26]:

$$\theta' = 2\pi(n \pm \frac{E}{E_c}). \tag{7.5}$$

This fact has important consequences. Although the amplitude acquires the Moyal phase as well, the modification of θ covers those features of the phase which in field theory are responsible for pathological acausal behavior at the "tree level," and failure of determinism at the "one-loop" level. In fact, in this case, the actual configuration of the physical system given by the outgoing wave function contains only time delay terms. The advanced terms are not there.

Broadly speaking, the acausal phase gets multiplied by a function generated by the oscillation effects of the string. What is left over by this multiplication is a formula in which the advanced terms do not appear because canceled. Therefore, the acausal behavior, generated by the non-commutative parameter θ at the two levels of perturbation, is removed.

Therefore, space-time non-commutativity does not compromise the deterministic structure of string theory and does not introduce pathological behaviors. The lack of pathological temporal features of strings physically explains why space-time non-commutative field theories cannot arise from low energy limits of string theory.

Finally, the appearance of a string's feature changing the non-commutative parameter as in (6.6) has a further important consequence. In fact, the formula clearly illustrates that despite removing the source of perturbation by turning off the electric field E, the non-commutative parameter does not vanish. Based on (6.6), time and space can exhibit non-commutative behavior also in the absence of a perturbation field. This fact does not seem to fit into the general picture presented by Seiberg, Susskind, and Toumbas, in which, as we saw, space-time non-commutativity holds just in that particular perturbative regime. This issue brings us to the content of the next section.

7.2 Space-time uncertainty principle in quantum string theory

Under what conditions does space-time non-commutativity arise in quantum string theory? Are we in the presence of some intrinsic feature of the theory instead of an extrinsic feature exhibited just in perturbative regimes? For answering these questions, I shall consider two different approaches.

The first view presented by Seiberg, Toumbas, and Susskind claims that String Theory shows space-time non-commutativity only in the presence of a background electric field.[27] The latter would produce a non-commutative perturbation of the theory in which a stringy space-time uncertainty principle can be consequently formulated. Its formal expression is a function of the α' string parameter[28]:

$$\triangle t \triangle x \geq \alpha'. \tag{7.6}$$

So, according to them, space-time uncertainty principle is an extrinsic principle derivable only in perturbative regimes of the theory.

A second view is that presented by Yoneya. In his essay, "String Theory and the Space-Time Uncertainty Principle" (see Yoneya, 2001), he claims that space-time non-commutativity is an intrinsic feature of the theory, which is not necessarily related to the presence of perturbative fields. More precisely, it arises from an intrinsic property of string dynamics. Therefore, it appears also in a non-perturbative conceptual framework of the theory.[29] The worldsheet intrinsic property from which it arises is connected in a special way with space-time uncertainty principle at the string length scale. Let's dig into this idea.

On the one side, according to Seiberg, Toumbas, and Susskind, the non-commutative relation $[t, x] = i\theta$ (valid only perturbatively) entails inside the same perturbative regime the spacetime uncertainty principle.

On the other side, according to Yoneya, things are the other way around: space-time uncertainty principle entails space-time non-commutativity. More clearly, the principle is thought to be one of those peculiar properties of the theory that can be derived from conformally invariant features of the string's worldsheet dynamics.

According to Yoneya, this principle is therefore "universally" valid.[30] More precisely, the uncertainty relation between space-like direction and time-like direction is just a particular expression derivable from a more general conformally invariant principle defined over the worldsheet. The latter, if read with respect to the Minkowski metric, produces the former.

The method of derivation of the uncertainty principle from worldsheet's properties is schematically presented in what follows.[31] The idea behind the derivation relies on the use of Riemann surfaces.[32]

Let's introduce some preliminary notions.[33]

An arc γ on a Riemann surface has a length $L(\gamma, \rho)$ with respect to a metric $ds = \rho(z, \bar{z})|dz|$. In general, $L(\gamma, \rho_1) \neq L(\gamma, \rho_2)$ if $\rho_1 \neq \rho_2$. However, it is possible to define "distances" on a Riemann surface which are conformally invariant, i.e. very peculiar "distances" that do not change if dilated or contracted. Let Ω be a region on the surface; let Γ be a set of arcs in that region; we can define the *extremal length* of the collection Γ of curves as a conformal invariant of Γ, which means that, given a conformal mappings $f : \Omega \to \Omega'$, the extremal length of Γ is equal to the extremal length of the image of Γ under f. The extremal length of Γ is defined as[34]

$$\lambda_\Omega(\Gamma) = sup \frac{L(\Gamma, \rho)^2}{A(\Gamma, \rho)}, \tag{7.7}$$

where $L(\Gamma, \rho) = inf_{\gamma \in \Gamma} L(\gamma, \rho)$ and

$$A(\Omega, \rho) = \int_{\Omega} \rho^2 dz d\bar{z} = \int_{\Omega} (\rho dz d\bar{z})^2 = \int_{\Omega} ds^2. \tag{7.8}$$

In particular, let "Ω be a quadrilateral segment"[35] (β, β' being the first couple of opposite edges and ξ, ξ' the second couple of opposite edges), and let "Γ be the set of all connected sets of arcs"[36] joining ξ and ξ'. Finally, let Γ^* be the set of all connected sets of arcs joining β and β'.

What is really important about the relation between these two sets of arcs is their *reciprocity relation*,[37]

$$\lambda_{\Omega}(\Gamma)\lambda_{\Omega}(\Gamma^*) = 1. \tag{7.9}$$

The conformal invariance of this relation comes from the conformal invariance of the extremal length $\lambda_{\Omega}(\Gamma)$.

Now, a particular history of a string, i.e., a string worldsheet, is represented by a Riemann surface – as Jeffrey Olson shows in his "Worldsheets, Riemann Surfaces, and Moduli."[38]
So, in particular, string worldsheets are conformally invariant surfaces on which we can define extremal lengths and their reciprocity relation. Path integrals are identified with maps from string worldsheets to a target spacetime.[39]

Now, the crucial thing to understand about derivation of space-time uncertainty principle is that the latter defined over the target spacetime of the path integral arises from the reciprocity relation between extremal lengths over the string worldsheet. This derivation can be understood looking at how a string's amplitude can be written inside the path integral.[40] In this amplitude, two extremal lengths $\lambda_{\Omega}(\Gamma)$ and $\lambda_{\Omega}(\Gamma^*)$ show up as the measure involved in probing spacetime's structure[41]:

$$exp^{-\frac{1}{l_s^2}\left(\frac{A^2}{\lambda(\Gamma)} + \frac{B^2}{\lambda(\Gamma^*)}\right)}. \tag{7.10}$$

Now,

$$\Delta A \sim \sqrt{\lambda(\Gamma)} l_s, \tag{7.11}$$

$$\Delta B \sim \sqrt{\lambda(\Gamma^*)} l_s. \tag{7.12}$$

Hence, the length l_s, the length probed by strings amplitudes in spacetime, can be expressed as

$$l_s \sim \frac{\Delta A}{\sqrt{\lambda(\Gamma)}} \sim \frac{\Delta B}{\sqrt{\lambda(\Gamma^*)}}. \tag{7.13}$$

Moreover, the extremal lengths $\lambda(\Gamma)$ and $\lambda(\Gamma^*)$ show up in ΔA and ΔB, respectively. Therefore, if we are examining both directions at the same time, the reciprocity principle involving those extremal lengths works as a constraint on how much information you can get at short distances. In fact, combining (4.10) with (4.12) and (4.13), we get

$$\frac{\Delta A}{l_s^2} \cdot \frac{\Delta B}{l_s^2} = 1, \tag{7.14}$$

which can be rewritten as

$$\Delta A \cdot \Delta B \geq l_s^2. \tag{7.15}$$

Interpreting what Yoneya says,[42] we can now read the above relation with respect to the Minkowski metric. If we do that, we will get a relation between a space-like direction and a time-like direction, which is exactly what Yoneya is trying to derive, i.e., space-time uncertainty principle at the length scale probed by strings.

Therefore, space-time uncertainty principle can be derived from the more general conformally invariant duality relation over the string's worldsheet. Then, according to this approach, the principle is an intrinsic property of string theory which in its turn entails space-time non-commutativity. Citing from Yoneya: "Thus the noncommutativity of space and time is indeed there in a hidden form.[...] What is in mind here is a different representation of string theory with manifest non-commutativity that is, however, equivalent at the level of the on-shell S-matrix, to the usual formulation."[43]

7.3 Conclusion

In the previous section, I considered two different ways of describing the conditions under which space and time fail to commute in string theory. According to Seiberg, Toumbas, and Susskind, space-time non-commutativity seems to be an extrinsic feature of string theory since space and time fail to commute only in some specific cases. According to Yoneya, although it is unquestionable that such perturbations yield space-time non-commutativity, they are not necessary conditions for the latter to happen. In fact, he shows that space-time non-commutativity can be alternatively derived from a conformally invariant principle defined over the string worldsheet. This fact qualifies space-time non-commutativity as an intrinsic principle of the theory, hence holding true in a non-perturbative formulation.

However, both views share the same idea that space-time non-commutativity does not raise the specter of indeterminism and acausal behavior. This fact has profound implications for the ordinary notion of space and time. The question is how the unbroken causality would support the emergent role of ordinary spacetime in string theory. As I said in the introduction of this chapter, the main idea is that spacetime might arise as an emergent structure from some underlying non-commutative structure that string theory might still account for, without losing consistency. Would this underlying structure be still geometrical?

Indeed, it would not. Rather, it would be much more similar to an abstract algebraic space, or equivalently a purely topological one devoid of any metrical property. Also, it should be characterized by some sort of discreteness and by non-locality.

Now, the profile of non-commutative "spacetime" presented in this chapter might satisfy these requirements.[44]

But then, in which sense would ordinary spacetime emerge from non-commutative "spacetime"? As we saw, non-commutative geometry contains the commutative one as its particular subcase obtainable by imposing a vanishing condition on the anticommutative parameter. So, ordinary commutative geometry would emerge from non-commutative geometry via the mathematical limit of $\theta \to 0$.

But in order to speak about the emergence of spacetime from non-commutative "spacetime," we need to figure out some kind of physical counterpart involved in that mathematical limit. Looking at (6.5), we can see that the non-commutative parameter θ and the physical parameter α' are directly proportional. So, the mathematical limit $\theta \to 0$ is formally connected to the physical low energy limit $\alpha' \to 0$. As we saw earlier on, the limit $\alpha' \to 0$ implements the physical low energy limit since in the theory, the only dimensionless parameter is $(\alpha' E^2)$.[45] Then, the low energy limit is what can characterize in a full physical sense this idea of an emergent spacetime from an underlying non-commutative "spacetime."

Now that we have a physical limit involved in the story of this emergence, we only need to understand whether string theory might still be the theory accounting for this non-commutative, topological structure. The theoretical findings we examined in the previous sections showed that introducing a non-commutative spacetime in a concrete case of string scattering does not break the unitary structure of quantum string theory. Space-time non-commutativity does not compromise the basic temporal features of quantum string dynamics. That can be read as the fact that emergent spacetime in string theory can be formulated via non-commutative geometry.

According to Yoneya, space-time non-commutativity (at the length scale probed by strings) is an intrinsic property arising from an underlying conformally invariant feature of the string worldsheet, namely, the reciprocity principle. Developing this train of thought, one may say that the underlying non-commutative structure, although more fundamental, is the ultimate one. Indeed, in this context, it would appear to arise from a deeper underlying worldsheet structure.

So, we seem to have a chain of emergent theoretical facts starting from the string worldsheet, where the reciprocity principle expresses a relation between "distances" peculiar to the strings world. Then, by discovering that space-time uncertainty relation is actually hidden in that principle, we can see that (still at the spacetime length scale probed by strings) a space-time non-commutative structure arises from that world-sheet by imposing particular constraints on the reciprocity principle. Finally, leaving the string length scale by a low energy limit taken on the non-commutative spacetime string theory, ordinary spacetime emerges.

Notes

1 This section develops some of the ideas presented in the fourth section of "Time in Quantum Gravity" – see Huggett, Vistarini, and Wüthrich (2013).
2 The Euclidean metric defined on a set of points is induced by the Euclidean scalar product "·" defined over the vector space underlying that set.
3 See Seiberg, Susskind, and Toumbas (2000a, 2000b) and also Gomis and Mehen (2000). Other bibliographic references can be found in what follows.
4 Gomis and Mehen (2000, page 1) and Seiberg, Susskind, and Toumbas (2000a, page 1.
5 Then, the ⋆-product between two fields is

$$\phi_1(x) \star \phi_2(x) = \exp^{\frac{i}{2}\theta^{\mu\nu}\frac{\partial}{\partial\alpha^\mu}\frac{\partial}{\partial\beta^\nu}}\phi_1(x+\alpha)\phi_2(x+\beta)|\alpha = \beta = 0.$$

6 Licht (2005, page 1).
7 Carroll et al. (2001, pages 2–3).
8 Also the rotation of a "localized field configuration."
9 An explanation of what unitarity means and how it relates to determinism will be shortly presented in the more important case of space-time non-commutativity.
10 Gomis and Mehen (2000, page 1).
11 Gomis and Mehen (2000, section 2) and Seiberg, Susskind, and Toumbas (2000a, pages 1, 2).
12 See Becker, Becker, and Schwartz (2007, page 301).
13 Gomis and Mehen (2000, section 2) and Seiberg, Susskind, and Toumbas (2000a, pages 1, 2).
14 For an exhaustive mathematical presentation of this problem, see Seiberg, Susskind, and Toumbas (2000b, pages 5–9).
15 S matrices are unitary matrices connecting asymptotic particle states, i.e., particle states between time equal to minus infinity and time equal to plus infinity; the entries of the matrix are scattering amplitudes which are produced by the interaction terms of the Lagrangian. These entries look like $< p_1, p_2|S|k_1, k_2 >$.
16 This amplitude is computed using Feynman diagrams. For more detail, see Veltman (1994, pages 46—62).
17 See Seiberg, Susskind, and Toumbas (2000a, 2000b, page 5, (2.26)).
18 See Seiberg, Susskind, and Toumbas (2000a, 2000b, page 6).
19 Seiberg, Susskind, and Toumbas (2000b, page 7).
20 Seiberg, Susskind, and Toumbas (2000b, page 8).
21 However, their reading of the "advanced" scattered wave packet as the space-like effect of the collision seems to be in tension with the fact that earlier on in the paper they seem to say that the "advanced" packet appears in the past cone of the collision.
22 See Gomis and Mehen (2000, page 2).
23 See Seiberg, Susskind, and Toumbas (2000b,page 9), Seiberg, Susskind, and Toumbas (2000a, pages 4–8), and Gomis and Mehen (2000, pages 9–12).
24 See Gomis and Mehen (2000, pages 10–11).

25 See Gomis and Mehen (2000, page 11, (3.3)).
26 I won't present here the mathematical steps relative to this point. For more detail, see Seiberg, Susskind, and Toumbas (2000a, 2000b, pages 11–12, formula 3.9).
27 See Seiberg, Susskind, and Toumbas (2000a, 2000b, section 3).
28 The space-time uncertainty principle tells us that when we try to probe short distances along a time-like direction, the amount of uncertainty relative to probing distances along a space-like direction increases and vice versa.
29 Yoneya (2001, pages 33–39).
30 See Yoneya (2001, page 7).
31 For more detail, see Yoneya (2001, section 2.3).
32 In mathematics, a Riemann surface is a complex manifold. It is usually represented as a deformation of the complex plane. A Riemann surface locally shares with the complex plane the same topological properties, whereas globally it appears to be topologically very different. A sphere is a good example of Riemann surface.
33 See Yoneya (2001, pages 11–13).
34 See Yoneya (2001, pages 11–13, (2.7)).
35 See Yoneya (1974, page 12).
36 See Yoneya (2001, page 12).
37 For an explanation of how this relation can be obtained, see Yoneya (2001, pages 11–12, (2.8)).
38 See Olson (2001, sections 2 and 3).
39 See also Yoneya (2001, page 11).
40 For more detail on how Yoneya derives this expression, see Yoneya (2001, pages 11–14).
41 See Yoneya (2001, page 13).
42 Yoneya (2001, page 13).
43 Yoneya (2001, page 33). Here, the equivalence at the level of the on-shell S-matrix means equivalence with respect to dynamics obeying the same equation of motion.
44 Let's say something more precise about non-commutative "spacetime." Non-commutative "spacetime" is an algebraic object arising from the extension of a commutative algebra of functions over an ordinary spacetime. Algebraic properties of a commutative algebra of functions perfectly encode the metrical properties of the manifold over which the functions are defined. For example, integrating over spacetime corresponds to computing the trace of some operator belonging to the algebra. So, in the commutative case, we have a perfect correspondence between the category of algebraic objects (algebra of functions over spacetime) and the category of geometrical objects (spacetime). However, when we extend the commutative algebra of functions over spacetime to a more general non-commutative algebra of operators – which includes the former as its particular case – we lose the correspondence between algebraic and metrical properties, because we end up with lacking a geometrical manifold over which the non-commutative algebra of operators is defined. So, non-commutative "spacetime" is identified with the non-commutative algebra of operators, and in this sense, it is an algebraic space.
45 See Becker, Becker, and Schwartz (2007, page 301).

Bibliography

David Albert. *After Physics*. Harvard University Press, 2016.

Paul Aspinwall. D-brane on Calabi-Yau manifolds. *arXiv:hep-th/0403166*, 2004.

Paul Audi. Grounding: Toward a theory of the in-virtue-of relation. *Journal of Philosophy*, 109:685–711, 2012.

John Baez and Javier Muniain. *Gauge Fields, Knots and Gravity*. World Scientific, 2008.

Vijay Balasubramanian, Jan de Boer, and Djordje Minic. Exploring de Sitter space and holography. *Classical and Quantum Gravity*, 19:5655–5700, 2002.

Vijay Balasubramanian and Per Kraus. Space-time and the holographic renormalization group. *Physical Review Letters*, 83:3605–3608, 1999.

Vijay Balasubramanian, Per Kraus, Albion Lawrence, and Sandip Trivedi. Holographic probes of anti-de Sitter space-times. *Physical Review*, D59, 1999.

Robert Batterman. *The Devil in the Details: Asymptotic Reasoning in Explanation, Reduction, and Emergence*. Oxford University Press, 2011.

Katrin Becker, Melanie Becker, and John Schwartz. *String Theory and M-theory. A Modern Introduction*. Cambridge University Press, 2007.

Mark Bedau. Weak emergence. *Noûs*, 11, 1997.

Jan de Boer, Erik Verlinde, and Herman Verlinde. On the holographic renormalization group. *JHEP*, 0008, (arXiv:hep-th/9912012), 2000.

Robert Brandemberger and Cumrun Vafa. Superstrings in the early universe. *Nuclear Physics*, B316:391–410, 1989.

Harvey Brown. *Physical Relativity: Space-Time Structure from a Dynamical Perspective*. Oxford University Press, 2005.

Jeremy Butterfield. Less is different: Emergence and reduction reconciled. *Foundations of Physics*, 41(6):1065–1135, 2011a.

Jeremy Butterfield. Emergence, reduction and supervenience: A varied landscape. *Foundations of Physics*, 41(6):920–960, 2011b.

J. Butterfield and J. Isham. Spacetime and the philosophical challenge of quantum gravity. in C. Callender and N. Huggett (eds.), *Physics Meets Philosophy at the Planck Scale*, Cambridge University Press, 2000.

Curtis Callan, Daniel Friedan, Emil Martinec and Malcolm Perry. Strings in background fields. *Nuclear Physics*, B262(4):593–609, 1985.

Philip Candelas and Xenia de laOssa. Moduli space of Calabi-Yau manifolds. *Nuclear Physics*, B355:455–481, 1991.

Sean Carroll, Jeffrey Harvey, Allan Kostelecký, Charles Lane, and Takemi Okamoto. Noncommutative Field Theory and Lorentz Violation. *arXiv:hep-th/0105082*, 2001.

Fabrice Correia. Grounding and truth-functions. *Logique et Analyse*, 53: 251–279, 2010.

Jin Dai, Robert Leigh, and Joseph Polchinski. New connections between string theories. *Modern Physics Letters*, A4:2073–2083, 1989.

Richard Dawid. *String Theory and the Scientific Method*. Cambridge University Press, 2013.

Dennis Dieks, Jeroen van Dongen and Sebastian de Haro. Emergence in holographic scenarios for gravity. *Studies in History and Philosophy of Modern Physics*, 52(Part B):203–216, 2015.

Michael Douglas and Nikita Nekrasov. Noncommutative field theory. *arXiv:hep-th/0106048v4*, 2001.

KIT Fine. The question of realism. *Philosophers' Imprint*, 1:1–30, 2001.

Jerry Fodor. Special sciences: or the disunity of science as a working hypothesis. *Synthese*, 28(2):97–115, 1974.

Bas C. van Frassen. *The Scientific Image*. Oxford University Press, Oxford, 1980.

Maurizio Gasperini. *Elements of String Cosmology*. Cambridge University Press, 2007.

Jaume Gomis and Thomas Mehen. Space-time noncommutative field theories and unitarity. *Nuclear Physics*, B591:265–276, 2000.

Brian Greene. *The Elegant Universe: Superstrings, Hidden Dimensions, and the Quest for the Ultimate Theory*. W.W. Norton & Company, 1999.

Brian Greene and Moshe Ronen Plesser. Duality in Calabi-Yau moduli space. *Nuclear Physics B*, 338:15–37, 1990.

Robin Hartshorne. *Algebraic Geometry*. Springer-Verlag, 1977.

bibitemHooft:2009 Gerard't Hooft. Dimensional reduction in quantum gravity. *arXiv:gr-qc/9310026*, 1993, last revised 2009.

Nick Huggett. Essay review: Physical relativity and understanding space-time. *Philosophy of Science*, 76, 2009.

Nick Huggett and Tiziana Vistarini. Deriving general relativity from string theory. *Philosophy of Science*, 82(5):1163–1174, 2015.

Nick Huggett, Tiziana Vistarini, and ChristianWüthrich. Time in quantum gravity. in Adrian Barton and Heather Dyke (eds.), *A Companion to the Philosophy of Time*, Wiley-Blackwell, pages 242–261, 2013.

Nick Huggett and Christian Wuthrich. Emergent spacetime and empirical (in)coherence. *Studies in History and Philosophy of Modern Physics*, 44(3):276–285, 2013.

Clifford Johnson. *D-branes*. Cambridge University Press, 2003.

Victor Kac. *Vertex Algebras for Beginners*. University Lectures Series, 10, 1997.

Gordon Kane. *Supersymmetry and Beyond: From the Higgs Boson to the New Physics*. Basic Books; Revised edition, 2013.

Immanuel Kant. *Critique of Pure Reason*, trans. by Paul Guyer and Allen W. Wood. Cambridge University Press, 1998.

Jaegwon Kim. Emergence: Core ideas and issues. *Synthese*, 151:347–354, 2006.

Kunihiko Kodaira. *Complex Manifolds and Deformation of Complex Structures*. Springer-Verlag, 1986.

Arthur Lewis Licht. A Lorentz covariant noncommutative geometry. *arXiv:hep-th/0512134v1 13 Dec 2005*, 2005.

Miao Li and Tamiaki Yoneya. Short-distance space-time structure and black holes in string theory: A short review of the present status. *arXiv:hep-th/9806240v1*, 1998.

Juan Maldacena. The large n limit of superconformal field theories and supergravity. *arXiv:hep-th/9711200*, 1998.

Juan Maldacena. The illusion of gravity. *Scientific American*, pages 56–63, 2005.

Marco Manetti. *Lectures on deformations of complex manifolds*. http://www.mat.uniroma1.it/people/manetti/dispense/defomanifold.pdf, 2004.

Brian McLaughlin. Emergence and supervenience. *Intellectica*, 2:25–43, 1997.

Ernest Nagel. *The Structure of Science: Problems in the Logic of Scientific Explanation*. Harcourt, Brace & World, 1961.

Timothy O'Connor and Hong Yu Wong. Emergent properties. *The Stanford Encyclopedia of Philosophy (Summer 2015 Edition), Edward N. Zalta (ed.), https://plato.stanford.edu/archives/sum2015/entries/properties-emergent/*, 2015.

Jeffrey Olson. *Worldsheets, Riemann surfaces, and moduli*. http://www.ph.utexas.edu/ jdolson/worldsheet.pdf, 2001.

Joseph Polchinski. Dirichlet-branes and Ramond-Ramond charges. *arXiv:hep-th/9510017*, 1995.

Joseph Polchinski. *String Theory, Superstring Theory and Beyond, vol. I*. Cambridge University Press, 2005.

Joseph Polchinski. Dualities. *Studies in History and Philosophy of Modern Physics*, 59: 6–20, 2015.

Karl Popper and John Eccles. *The Self and Its Brain: An Argument for Interactionism*. Springer International, 1977.

Dean Rickles. A Philosopher Looks at String Dualities. *Elsevier*, 2010.

Dean Rickles. AdS/CFT duality and the Emergence of Spacetime. *Studies in History and Philosophy of Modern Physics*, special issue on the Emergence of Spacetime in Quantum Gravity (edited by Nick Huggett and Chris Wuthrich), 2012.

Gideon Rosen. Metaphysical dependence: Grounding and reduction. in R. Hale and A. Hoffman (eds.), *Modality: Metaphysics, Logic, and Epistemology*, Oxford University Press, pages 109–136, 2010.

Joel Scherk. An introduction to the theory of duals models and strings. *Reviews Modern Physics*, 47:123, 1975.

Joel Scherk and John Schwarz. Dual models for non-hadrons. *Nuclear Physics B*, 81:118–144, 1974.

Nathan Seiberg. Emergent spacetime. *arXiv:hep-th/0601234*, 2006.

Nathan Seiberg, Leonard Susskind, and Nicolaos Toumbas. Strings in background electric field, space/time noncommutativity and a new non critical string theory. *Journal of High Energy Physics*, 06(06), 2000a.

Nathan Seiberg, Leonard Susskind, and Nicolaos Toumbas. Space/time noncommutativity and causality. *Journal of High Energy Physics*, 0006(06), 2000b.

Laurence Sklar *Space, Time and Spacetime*. University of California Press, 1977.

Mark Srednicki. *Quantum Field Theory*. Cambridge University Press, 2007.

Andrew Strominger. The dS/CFT correspondence. *JHEP*, 0110:034, 2001.

Washington Taylor and Barton Zwiebach. D-Branes, Tachyons, and String Field Theory. *arXiv:hep-th/0311017*, 2004.

Nick Teh. Holography and emergence. *Studies in History and Philosophy of Modern Physics*, 44(3):300–311, 2013.

David Tong. Lectures on string theory. *arXiv:hep-th/0908.0333*, 2012.

Cumrun Vafa. Geometric physics. *Proceedings of the International Congress of Mathematics*, 1:537–556, 1998.

Martinus Veltman. *Diagrammatica. The Path to Feynman Diagrams*. Cambridge University Press, 1994.

Gabriele Veneziano. Construction of a crossing-symmetric, Regge behaved amplitude for linearly rising trajectories. *Nuovo Cimento*, A57:190–197, 1968.

Tiziana Vistarini. Holographic space and time: Emergent in what sense?. *Studies in History and Philosophy of Modern Physics*, 59:126–135, 2016.

Marcel Vonk. A minicourse on topological strings. *arXiv:hep-th/0504147*, 2005.

Edward Witten. Mirror manifolds and topological field theory. *arXiv:hep-th/9112056v1*, 1991.

Edward Witten. Quantum background independence in string theory. *arXiv:hep-th/9306122v1*, 1993.

Edward Witten. Reflections on the fate of space-time. *Physics Today*, 49N4:24, 1996.

H.Y. Wong. The secret lives of emergents. in Antonella Corradini and Timothy O'Connor (eds.), *Emergence in Science and Philosophy*, Routledge, 2010.

Tamiaki Yoneya. Connection of dual models to electrodynamics and gravydynamics. *Progress of Theoretical Physics*, 51:1907–1920, 1974.

Tamiaki Yoneya. String theory and the space-time uncertainty principle. *arXiv:hep-th/0004074v6 30 Mar 2001*.

Barton Zwiebach. *A First Course in String Theory*. Cambridge University Press, 2nd edition, 2009.

Index

For Product Safety Concerns and Information please contact our EU
representative GPSR@taylorandfrancis.com
Taylor & Francis Verlag GmbH, Kaufingerstraße 24, 80331 München, Germany

www.ingramcontent.com/pod-product-compliance
Ingram Content Group UK Ltd.
Pitfield, Milton Keynes, MK11 3LW, UK
UKHW021437080625
459435UK00011B/291

* 9 7 8 1 0 3 2 1 7 7 9 2 2 *